Other Books by
Francis Raymond Line

coauthored with Helen E. Line

GRAND CANYON LOVE STORY

Published 1984
Wide Horizons Press

MAN WITH A SONG

Major and Minor Notes
In the Life of
Francis of Assisi
Published 1982 by Image Books
A Division of Doubleday & Company, Inc.

BLUEPRINT FOR LIVING

Published 1977 by
The Upper Room

SHEEP, STARS, AND SOLITUDE

Adventure Saga of a Wilderness Trail

by
Francis Raymond Line

Published by

WIDE HORIZONS PRESS
13 Meadowsweet
Irvine, CA 92715

Library of Congress Cataloging in Publication Data

Line, Francis R.
   Sheep, stars, and solitude.

   1. Lucero, Rosalio.   2. Shepherds — Arizona — Biography.
3. Sheep — Arizona.   4. Heber-Reno Stock Trail (Ariz.)
I. Title.
SF375.32.L83L56   1986   636.3'0092'4   86-11154
ISBN: 0-938109-02-2 (pbk.)

To
Helen Gibson Line
a Native of Arizona
who helped me fall in love with her state

She was raised on an Arizona cattle ranch,
one of eight daughters,
the "Gibson Girls" of Globe

Her parents came to Arizona by covered wagon
from Texas and New Mexico

They helped pioneer the West
Helen has helped me portray it.

Helen Gibson Line

Thanks

—to the editors of NATIONAL GEOGRAPHIC MAGA-ZINE and ARIZONA HIGHWAYS, who used my original stories and pictures of the great sheep trek and whose encouragement helped lead to the writing of this book.

—to ENCYCLOPAEDIA BRITANNICA EDUCATIONAL CORPORATION for distributing throughout America, during these many years, my color motion pictures of this adventure.

—to DAN GENUNG, whose Arizona heritage began in 1863, before the area became a territory, for reading and approving the manuscript. And for writing the Foreword to this book.

—to the HERD OWNERS, past and present, without whom there would be no sheep treks in Arizona — Irving and Margaret Gibson, the Dobsons, Gunnar Thude, Elma Sanudo, Bill Ryan and all the others.

—and especially,

—to the unsung heroes of the Arizona sheep treks — the HERDERS and CAMPEROS, among them being Rosalio and Pablo, my friends and companions of the trail.

Francis Raymond Line.

Contents

Episodes of the Trail

Foreword

Some comments about this book by a third generation Arizonan
Dan Genung.

When you open the pages of this book, plan to live for several hours with life at its simplest, harshest, and yet strangely its most gloriously beautiful. You will be living with history on the hoof.

You must be willing to march with the planet Jupiter as it leads the parade of night across Arizona skies. You learn to seek sleep in thin blankets while the stars brush your hair and the Milky Way nestles on your shoulders.

Here is your challenge to trudge through searing sands and on rocks that shred your shoes. You will be trapped on a mountain slope which has become a "rock-ribbed refrigerator" by the freezing impact of a storm "like a Paul Bunyan gone mad in the forest."

Here is your invitation to dine on beans and pan-brown bread, to share the misery of a shivering sheepdog and the rain-sodden woes of burros who "document their sorrow with their ears at half-mast."

Here is your introduction to Rosalio Lucero, whose last name means "Morning Star," a sheepherder whose profession predates Moses by thousands of years, and whose unfolding odyssey seems to translate the scenes of the Twenty-third Psalm into a setting of the Arizona wilds. You can share his heartaches and joys as he nurses 2,000 "sheepsies" over 200 treacherous miles "with the tender care of a mother with a colicky child."

The writer of this true adventure is Francis Raymond Line. When I first met him, he struck me as a five-foot bundle of startling strength and energy. Now, 28 years later, I see him as six feet eight in height, a man with a magic camera, the soul of a prophet, and the word-imagery of a Walt Whitman. See for yourself!

11

When Francis was eighteen, he and his brother hiked through the forty-eight states. Lectures and films on his travels helped him earn his way through the University of Michigan. In the sixty years since then his film presentations and talks have thrilled audiences from Carnegie Hall to Hawaii, from Canada to the Gulf of Mexico. His most popular film, SHEEP, STARS, AND SOLITUDE, the background of this book, has been translated by the U.S. State Department into two dozen languages. A shorter version, MORNING STAR, circulates through film libraries and schools and colleges of the land.

Two of his best-known books written jointly with his wife, Helen Line, are MAN WITH A SONG, based on the contemporary impact of St. Francis of Assisi, and GRAND CANYON LOVE STORY. Francis and Helen spent days exploring the Grand Canyon the first years of their married life, hiked it many times, and celebrated their Golden Wedding anniversary and each year since by hiking to the bottom for an overnight stay.

As you read these pages you will be blistered by heat, battered by hailstones, stabbed by cholla cactus, but above all thrilled at the faithfulness of Rosalio, a heroic human whose history-laden tasks are a ministry of love. You will know the healing power of Stars and Solitude.

Dan B Genung

An Introduction to the Trek

My first journey along one of America's roughest and wildest sheep trails — the Heber-Reno of Arizona — was for the purpose of producing a color motion picture with which to lecture on travel-adventure courses and in schools and universities across the country. I made the journey a second time to take still photos and write articles for The *National Geographic Magazine* and *Arizona Highways*. Each time I followed sheep in charge of Rosalio Lucero, the herder, and Pablo Chavez, his campero.

The first trek took forty days. On the second trek we journeyed for fifty-two days, Rosalio with his herd of 2,200 sheep, Pablo with his seven burros to carry our bedrolls and supplies, and I on foot to catch as best I could the warm breath of the trail, the smell of the wilds, and the story of the sheep. I was not an outsider; only an extra bedroll, and that a thin one, was taken along, and a few more cans of tomatoes and pears. I became a part of the migration.

In the elapsed years I have from time to time come again — often with my wife and daughter — to meet the herds once more at the occasional watering places, to renew the fragrance, to recapture the lure of the trail. Each fall and winter I carried my film of this adventure to cities of the Midwest and East and have tried to bring to the people in those packed auditoriums a breath of the wilds. But the halls are hot with the breaths of people instead of with wildness, and the lecture hour is short. There is much, too much, that cannot be said in pictures, about the lore and adventures of the Heber-Reno, and about the qualities of Rosalio, who elevated sheepherding into heroic craftsmanship.

My concepts of nature, because of Rosalio's influence, expanded profoundly, as well as my ability to overcome fears. In a sense, the sheep trail became my Walden — a fifty-two day opportunity for retreat, where lessons in nature were taught and demonstrated on a daily basis.

Because of the unusual quality of its subject matter, my color motion picture of the sheep drive became the most popular adventure program on the American illustrated lecture platform. (See following page.)

I tell the story as it unfolded, on my second trek, in mid-twentieth century Arizona, before changes and "civilized" modifications began dispelling the grizzly grandeur. The chronicle is important as a historical record. I tell it so that the story of Rosalio and the Heber-Reno Stock Trail may be preserved, as an inspiring contribution to authentic Americana, and to the lore of Arizona and the Southwest.

Enjoy the journey.

Francis Raymond Line.

James Pond, founder and editor of PROGRAM MAGAZINE, the "bible" of the illustrated lecture world, wrote a special editorial after seeing Line's color motion picture, SHEEP, STARS, AND SOLITUDE, the basis for this book.

We Bow to Line

*I*f PROGRAM had Hollywood Oscars to hand out, it would give an award to Francis R. Line for his motion picture epic, "Sheep, Stars, and Solitude." Forgive the use of the adjective "epic." But this is an epic in every sense of the word. Not only would PROGRAM make this award, so would his fellow lecturers in the motion-picture field. When Francis Line was recently in New York his conferes went from one lecture of his to another, to drink in while they could the fragile quality of this film of loveliness. They talked about it for days after Line had gone, trying to analyze why this film was in a class by itself. They analyzed it to the nth degree, and then finally concluded that it was great because it had an idea. One might almost say a soul.

What is it about? Just what the title says. Sheep, Stars, and Solitude. It is the story in moving color of the migration of herds of sheep from the desert regions of Arizona, where summer heat would be unendurable, to cool mountain regions where lush meadows await them.

But what an ordeal for the sheep. And their patient shepherds who must walk with them. The trek takes forty days and forty nights, which reminds us of another Shepherd and an agony of forty days. The sheep must cross waterless desert areas, climb almost impassable mountains, break trail through trackless scrub and brush. Their feet get injured. Their souls grow weary. But the shepherds nurse them along.

Francis Line's filming of the journey is perfection. Every inch is beauty. Every foot carries a heart-beat. With a dash of welcome humor.

Pictures like this restore the soul. No praise too high, no award too great, for the man who has produced this, the perfect film.

15

Map, Courtesy The National Geographic Magazine

CHAPTER 1

Rosalio

Rosalio Lucero is a hero. He does not remotely resemble one, yet twice a year he performs a hero's task guiding his migrating herd over the tortures of the Heber-Reno Stock Trail.

I had to see him experience heaven and hell for fifty-two days, as we followed one of the wildest sheep trails in America, before I comprehended the humble qualities of his greatness. He is a Mexican-American sheepherder who, for much of his life, has been driving herds each spring on the long trek from Arizona's hot central valley to cool summer pastures in the mountains, returning with them each autumn.

His furrowed tanned face is a relief map of his Arizona hills. The brown skin of his hands, and particularly of his cheeks, and about his eyes, is ravined like the hard but wildly grand terrain of the herd country. At three score years his hair is still as black as the charcoal embers of his campfires. His teeth — although not brushed with any regularity, yet preserved and hardened because of simple eating habits — are white as bleached bones on the desert. Forty-seven years of leaning, in odd moments, on a herder's staff have sculptured his shoulders into the curve of the staff itself.

Rosalio dresses in dusty bib overalls and denim jacket which flap grotesquely as he runs in pursuit of an errant sheep.

On government maps this trail and its extensions are shown by a neatly patterned arc-shaped crimson line connecting a pinpoint near Phoenix with another pinpoint beyond McNary, Arizona, up toward the New Mexican border.

To the sheep, trekking northward each spring to escape the

17

heat of Phoenix's Salt River Valley, returning each fall to escape the chilling snows of the White Mountains near McNary, the trail is a blessing tinged with blood.

As the crow flies — although hawks, vultures, and even eagles are the winged surveyors which more commonly measure distance over these cactus-calloused deserts — the length of the trail is over 200 miles. The sheep make it at least 300, meandering as they do like streams of water to seek passage over wild terrain. For more than fifty days every spring, because of the economy of the Arizona sheep business, plus the stern commands of climate in the state, the sheep must struggle upward to the mountains, only to return in the fall. They and the herders are trailbound nearly a third of every year.

Rosalio, the herder, who must keep the sheep in progress and circle constantly in search of strays, travels a far greater distance than the sheep. On a round trip in spring and fall he covers probably 700 miles, on foot all the way, nearly every step a strenuous climbing one, and a test of strength.

The Heber-Reno is a long, almost pathless strip from two to four miles wide laid out by the government as a driveway for the herds going from central to east central Arizona. Established with almost no regard for terrain, it plunges into ravines, surmounts mountains, and penetrates forests of cactus and of pines.

The trail embraces the whole range of seasons. The herders traveling north each spring find that debilitating summer soon stems into refreshing spring, sometimes winter, as the trip progresses. The calendar of half a year is condensed into fifty days.

It spans the range of wild life and things which grow. Brown lizards, sunbathing in lethargy on rocky deserts, give way to brown deer, bouncing gloriously in silent forests. The towering cactus dissolves into towering pine. There are many trails in America where more sheep travel, but no other with the variety of this one.

Historically the Heber-Reno Trail accomplishes the mightiest span of all. For time dips backward swiftly as each day carries the sheep away from signs of civilization and on into silent hills; time recedes until those hills of Arizona become closely knit in a historical fleecy skein with the hills of Babylonia and Palestine, where sheep have provided the needs of humanity and woven themselves into the lore of the land since earliest biblical times.

The sheep and their herder can go for forty days on this trail without trespassing on private lands. There are fifty miles between fences. Only a few times during the first month are roads encountered. Not one cactus in that month of travel is owned by any individual, and each lizard which dwells there is accountable only to God and the predators. The whole business, save for a short patch near the end and a little stretch which goes through Indian territory, is public domain. The Heber-Reno is an undulating ribbon of wildness.

CHAPTER 2

Destiny and Dust

A handful of ranchers, most of them with homes in the Tempe-Mesa-Chandler triangle of the Salt River Valley, and with summer ranges in the northern forests, dominate the sheep business of the Heber-Reno Stock Trail.

The number of sheep involved is not large; it is the manner of handling and the terrain of the trek which scratches one's fancy. A quarter of a million wool-bearers make the annual migrations but two-thirds of them take the Ashfork Trail through Lonesome Valley, or the Flagstaff route through Bloody Basin, or the Beaverhead-Grief Hill Route.

Only 30,000 — and the number is decreasing until one day the whole thing will be jelled and packaged in history, for the figure used to be 230,000 — followed the Heber-Reno Trail at the time of my second journey.

The Salt River Valley surrounding Phoenix is a winter resort not only for people but for the herds as well. For six months — from October to April — the sheep breakfast and sup in alfalfa which drips of greenness, a herd of 3,000 scouring out a pasture in a week, at high rental cost to the owner.

By mid-April the scorched valleys of Old Mexico have started blowing their hot breaths northward across the border. Most of the greenness has been sucked by the sheep from the pastures of alfalfa. Residents of the Salt River Valley turn on their air coolers, but for the sheep the only air-conditioning is migration. The signs of the seasons in the valley, the zodiacal signs in the sky, are all arguing for the northbound trek to begin. In mid-April the luxury hotels of the area prepare to bar their doors for the summer, plow up the lawns and the gardens to be replanted in the fall, and send their herds of relaxing millionaires back east till another season.

By mid-April the four-footed sheep have likewise been shorn, crews coming in with power clippers to strip off the wool before the hot journey begins.

From October to April Rosalio and Pablo had tented in the pastures, living like the Arabs of old. During the last three weeks they had thrown themselves unstintingly into preparations for the journey; pack boxes painted a lusty red, hobbles and harness mended, Pablo's horse shod (Rosalio would walk all the distance), their own shoes stoutly soled, the pack boxes concentrated with provisions. Paraphernalia littered the campsite as though this were an exodus of all the children of Israel from Egypt. When the seventh burro was firmly packed — 150 pounds to some of the animals — there was enough remaining to weigh down another animal. But by judicious planning this excess was roped and tied and hawsered to various of the seven — shovels and axes here, Dutch ovens and lantern there.

April 16. The sheep under Rosalio's care were poised to depart from the winter pastures. Gunnar Thude, the owner, who would freshen the herders with supplies along the trail, and a couple of Mexican ranch helpers, were on hand to assist at the start.

Four of the Dobson herds had broken camp and started on the 15th. All during that day four frothy kites of spiraling dust marked their progress along the dirt roads across the valley. And now, with the morning sun of another day bombarding over Superstition Mountain and firing its hot shafts horizontally across the pastures, the dust of Gunnar Thude's first herd, which broke camp before daylight with Mulatto as herder, became a puffball of rose-colored cotton as it drifted along the valley.

Rosalio and Pablo, marshaling Gunnar Thude's second herd, with 2,260 sheep in the band, left camp after daylight.

As the pasture gate was opened and sheep began streaming onto the road a melody of baaing arose, puncturing up into the vast balloon of dust. The river of sheep, packing the roadway from

22

fence to fence and washing and slopping over into the fields, flowed like a spring freshet released from winter thaws.

Ahead of the woolen river flowed the clouds of dust and ahead of that, as though carried on some fresher current, flowed the swelling baas of the herd.

The sound of the sheep leaving the pastures reached out over the fields and on across the valley to the town of Chandler nearby. The Arab delegates to the United Nations Assembly in New York — desert sheiks in silk robes who had come westward to taste the desert, Arizona style — those Arab princes who were just departing from their hotel, might have heard that swelling melody of the sheep, that trail song of the herd, and carried it in memory back to Araby.

If so they would have returned the echo to its origin. For this baaing of the Arizona sheep was simply a repeated chorus of the sheep sounds which had filled the mornings in Arabia when the sheiks draped themselves, not in silk, but in lambskins, and quieted their hunger with mutton.

There has been huge progress since Scheherazade wove her yarns of the Arabian Nights and sheep grazed the hills of Araby. Now we ride the Constellation instead of a camel. The siren replaces the lute. The knife plunged to the hilt in slaughter has a uranium edge. Both comforts and confusions have been multiplied.

Through it all, only the sheep have been constant, a familiar link in the chain of history. For sheep do not fly; they plod. They walk at a mile or so per hour. To the degree that people are still dependent on the sheep, humans likewise plod, and in these rare moments their feet are kept on the ground.

We are still dependent on mutton and its newer derivatives for our meat; much of our warmest raiment still is spun from the wool; and milk — from sheep, or goat, or the slightly more modern cow — is yet the calcium of our existence. We pasteurize the milk, and aerate it, solarize it, homogenize it, and add Vitamin D. But it still comes from the teats of the herd. And the symbol of a higher education — but whether this is slander to the herd, who can say? — through all the years has been the parchment sheepskin. This was an elemental river of destiny flowing along the dusty Chandler roads, a stream of meat and milk and raiment.

History on the hoof.

Above the herd flowed the dust. For two days the undammed stream of sheep spewed the powdery offal of the roadways into the air, stopping only at night, striving to finish this first test of the journey. Behind the herd, Rosalio drank dust all day as he climbed fences on one side of the road and then the other, urging on the stragglers, those little pools of sheep which eddied out of the main stream into the fields and would otherwise have become stranded, like stagnant water holes.

At night he coughed from the dust in his throat, and next day he drank it again.

Gunnar's pickup truck came by and I took a ride in it. Working our way through the sheep on the road was like plunging a wedge through a bale of wool. The firmly packed herd oozed slightly on either side and somehow made way for our passage. We were moving on a wide gray cloud. For fifteen minutes we wedged our way through and by then the sway of movement had become a surging, foamy sea, with a low hanging fog of dust hindering our vision.

That second afternoon we crossed the final irrigation canal with sides so precipitous the sheep could not drink. On trucks from a convenient loading place Pablo and Gunnar brought water in barrels for the burros and Pablo's horse.

Gunnar turned back. Leaving the roadway after crossing the final canal, the herd pushed on. The dust of this first stretch of open country was different from the dust of the road. The stream of the herd could flow wider, the dust cloud became thinner and broader.

Soon we crossed U.S. Highway 60 with its blatant signs and crude structures. As the herd struck across this final main thoroughfare, as though glad to be shed of these trashy tag ends of civilization, it vomited up a last great cough of desert dust which choked the throats of the auto travelers who had stopped to let the sheep go by. The dusty cloud hung in the sky for half an hour, its particles finally catching the rose of the sunset as, windswept at last, they drifted away and left the desert air clear for the stars.

CHAPTER 3

Sheep Bridge

The true trek begins on the third day at the crossing of the Salt River, at a swift-flowing section near Blue Point. The crossing provides a clattering opening drama. In former times the herds used to swim the river but this was not only a difficult task for animals and herders but dangerous and costly as well, with occasional drowning of sheep and burros. So the Arizona Wool Growers Association and the Federal Government jointly constructed a narrow suspension bridge, 320 feet long with its approaches, which in true western style is spoken of as "the only bridge in the world exclusively for the use of sheep." There are other sheep spans, and furthermore this one is for burros, dogs and herders, as well as woollies. Yet, even though it may have rivals, the Salt River sheep span affords concentrated drama. (This bridge has subsequently been destroyed by flooding of the Salt River. The herds now cross on a road bridge close by.)

Our herd swung down to the river close by the bridge at mid-morning and Pablo at once laid out his camp and began cooking in restaurant quantities. For this river crossing much resembles old-fashioned threshing day on a midwest farm; there are extra mouths to feed.

Before noon the owner of the herd, accompanied by his wife and daughter, bumped in with their pickup truck down the sandy winding road leading through the brush from the main dirt highway. The pickup was stocked with canned goods for the men, and salt for the sheep. This bridge is one of the few places en route where additional supplies can be brought in.

The government ranger pulled in shortly afterward. Much of the trek is over lands of the Tonto National Forest and a fee per head is charged by the forest service. This crossing of the river, as the animals stream off the bridge, is a heaven-sent opportunity for obtaining a correct official government tally.

The herd of another sheep outfit had also arrived for the crossing. Though the animals looked like all the rest, they were different; they were encumbered with a mortgage. A man from the finance company was on hand to tally his mobile collateral as it surged across the span. This sheep bridge was a countinghouse in both the figurative and literal sense.

The campero from the other outfit brought some larder to aid Pablo and at noontime all of us — the two camps thrown together, along with owners, ranger, mortgage-holder, and guests — filled our plates with frijoles and mutton shanks and ate pan biscuits which Pablo had baked in good quantity. For an hour afterward we wove woolly yarns about the lore of the trail and of the herds. A day at the sheep bridge is a combined supply bivouac, financial function, government survey, and a carnival occasion. It is a farewell fling which the herders enjoy before they plunge into fifty days of wilderness.

And then the crossing. Our herd went first. The animals were bunched into a corral until they resembled a moving sea of gray foam. A gate was opened onto a chute leading to the bridge, there was a moment of hesitation until the leaders were started, and the excitement began. For spectators it was sheer delight. The bridge, light in weight and of the suspension type, writhed and trembled as the sheep pounded across. Its gyrations sent fear into every animal and they ran and pushed and stormed to get over. Only three feet wide and intended for single file passage, the bridge under duress often must accommodate the sheep two abreast, since they are in such frantic haste to be done with it.

It is common for sheep to leap into the air as they are tallied. But as they reached the end of the bridge, where the ranger was making his census, they beheld solid ground again and made enormous jumps of joy which hurtled them high through space like circus animals vaulting through hoops.

I took a try at the tally but in five minutes I was dizzy and had lost my count. Amateurs sometimes faint endeavoring to count sheep in this fashion, the herd flashes past so swiftly. Each animal is branded with the owner's insignia painted or stamped on its back, and these large brands flashing by accentuated the dizziness that affects anyone making the count. For most persons, sheep counting is best done in somnolent moments only, and always in imagination. Only the expert does it well in daylight with his eyes wide open.

For nearly half an hour the stream of woolly backs rushed past, hooves pounding, air electric with the bleating, and with those final circus leaps at the end. It was a lively prologue to the trip.

The herders struggled for another half hour trying to induce the burros to follow across, for they would freeze in their tracks rather than run, as the bridge began to sway. The ranger made out his trail permit for Rosalio, the sheep were driven down to the river for one last drink, good-byes filled the air all around and soon we and the herd disappeared from the sight of those remaining at the bridge.

CHAPTER 4

Generation Gap

Rivers have always been boundary lines, marking changes which are either political or geographic or commercial. Seldom are two sides of any great river the same, and in some instances the change is compelling. The Salt River at the sheep crossing is such a type.

As we left the bridge, the sound of voices, the confusion of the crossing, the blasphemy which had spurred the burros to their efforts, all slipped away as easily as though a dirty jacket were being removed from this part of the world. The river itself dropped from view and a scar-faced mountain ahead became our new horizon.

The dust remained. As the white glare of afternoon dissolved into the quiet yellow of evening, then shifted to orange, the dust cloud absorbed each new tint, clutching and holding it for several moments, as though jealous of every different hue. Then we slipped into a narrowing canyon which, without any warning, squeezed all the color in an instant from the dust cloud. It was evening.

Generally only the ewes (or sometimes yearlings which will be mothers by winter) make the Heber-Reno journey. Lambs, born in the winter pastures, are sold before the spring trek commences, except for a few which may be taken along so they can learn the trail for future years.

No bucks (the term used by Arizona herders) undertake the arduous trek. These pampered creatures — some sixty-five of them for each 2,000 ewes — are transported hundreds of miles by truck and by rail, reaching summer pasturage in the north by a long circuitous route which leaves them fresh and rested. If breeding took place en route, lambing would occur under undesirable conditions before the herd returned to winter pasture in the fall. But another reason is that the bucks, because of their breeding equipment, cannot endure the hardships and hazards of the trail. That is left to the ewes.

In the corral back at the bridge, Gunnar Thude had caught several sick or lame ewes and lambs, to be returned to the ranch. But one lame lamb had somehow evaded detection.

It was the last sheep into the darkening canyon. Frightened at its rear position and endeavoring in a frantic run to catch up, it tried to take a shortcut course between two boulders. The space was too narrow; the creature wedged.

Circling constantly, flanking one side and then the other, dropping back to retrieve stragglers, Rosalio even in the semidarkness discovered the stranded animal quickly. With stout jerks he set it free.

Fifteen minutes later the lamb, still behind, lay down and refused to move at all in spite of Rosalio's urgings. The herd, braced by the cool air which the inflowing darkness had released into the canyon, was traveling fast. The herder administered everything at his command but the lamb lay back. To leave it was the only choice. The creature's breath came fast and its eyes showed fright as it watched us follow on behind the herd into the shadows ahead. The canyon walls soon cut off its cries. "It's bad," Rosalio explained to me in broken English. "The lambsie no should have come."

Twilight was already upon us. Rosalio began touching fire to the cholla cactus. I wondered why, but the occasion for an explanation did not present itself until next day. Each cholla roared into flame like an isolated candle and the effect in the still, vast desert was ghoulish but beautiful.

There was no feed and the sheep maintained their swift pace. But this herd which Gunnar Thude had put into Rosalio's charge was a mixed congregation. Most of them were old ewes which had been this way before, two, three, and four times; some had made as many as seven pilgrimages back and forth to the mountain Mecca. Sheep remember well. The old ewes knew the trail. But the rest —

30

the yearlings raised in the valley — had never passed this way.

Thrown together, the two age groups concocted for Rosalio a problem as old as civilization. Youth must lead but does not know how. The ewes, naturally the leaders, were slowed by weight and age. The yearlings, strangers to the desert and to the trail, snatched the head position and wandered away into unfavorable canyons. Rosalio pursued. I saw his flapping faded jacket making a climb out of a ravine to the left. A few minutes later it was flapping off there to the right, as he chased after a band of yearlings.

When next Rosalio came my way he pointed out a dark hill ahead where Pablo — gone on with the burros — would be making camp. Leaving the herd, I headed toward the hill, and soon came upon the campero.

Pablo was clearing the brush for a suitable campsite. Quickly and skillfully his burros were unpacked and hobbled, the kyack or pack boxes arranged methodically. Two pack boxes covered with an old square of canvas made a work table. It was dark and he asked me to build a fire as a beacon for Rosalio.

As preparations for supper were started, we heard bells toward the southwest. Some of the sheep wear bells to help Rosalio keep track of the herd and make sure none is lost. By the time I had gathered a great loose pile of wood — mainly ocotillo skeletons, for other kinds were scarce — Rosalio and the sheep came up.

Pablo had picked his camp to take advantage of a wood supply and a breeze. The desert sloped away toward the river and a fresh wind flowed up toward us from Granite Reef, which was sawed out of the sky-glow where the sun had set. The wind blew the world clean, carrying with it the heat pockets, the stale air, the dust of the day. It was late, so some canned corn, warmed-up frijoles, and the remaining biscuits from noontime — all washed down with coffee — made our meal. For dessert there was sorghum syrup which the herders ate "as is" but which I poured over the biscuits.

But there was no supper for the sheep — only scattered bunches of wild grass which had dried down to its roots. The herd, coming up to the camp, had split on either side of us and slowly drifted on past, up a long low hill to the east. They are loathe to stop at night when forage fails.

Rosalio had time to roll out his bed and make a start with his supper then, taking a flashlight, he followed the sheep, for they must be stopped and bedded down before the day for him could close. Pablo and I waited for half an hour, watching the beam of the flashlight bob along the low borders of the hill, then in the darkness zigzag far away toward the top. It was rough up there. "He can't stop them," said Pablo. "He's trying to bed them down. He'll be a long time coming." He *was* a long time coming — more than an hour after darkness had settled. It was 8:30 when Rosalio returned to supper.

I did not wait to lend a hand with the dishes — my usual chore. This first real sampling of the trail had battered my feet even through heavy shoes, for this desert floor is a grindstone set sideways. As I was slipping to sleep there came a sharp bay of coyotes back along the trail, and their call told clearly what had happened there. The little lamb, left alone in the black night of the canyon, would never be afraid any more.

A pair of coyotes echoed from the north with a series of confused and peculiar yelps. "That means rain," Pablo observed. "I been on this trail a long time and the coyotes always right. We have a change of weather tomorrow. You see."

CHAPTER 5

Longest Day

Pablo was building the fire when I wakened before dawn. The last star of the Dipper handle had arched about until it was nearly overhead. The sky of early morning is a different sky than evening, actually a preview of summer's major billing to come. It is the second feature of a double bill which few people are awake to see.

We each cooked our own eggs. The boss had supplied us with a couple of dozen at the bridge and I shall always marvel at the manner in which Pablo can transport eggs burroback over some of the roughest terrain in America, without any casualties. Nesting them in his tall boxes of oatmeal and flour, he never breaks a one. Balancing out the eggs with coffee, pan bread and stewed fruit, we were packing camp before daylight. Rosalio was off with the sheep by the time the gray light had come. Rolling up my own blankets but leaving them for Pablo to pack on the burros, I followed.

Half an hour later the desert plants about me — the cholla, saguaro, ocotillo, and palo verde — began to green and brighten as if some strangely moistened earth were sending new life into it. That is the way the sunrise came this morning, not a burst of whiteness over the hilltop but like a seeping upward of light from the roots of the desert growth. The coyotes and Pablo had prophesied correctly; the weather had changed. The sky was finely webbed with misty clouds, which caught and restrained the light.

With sunrise came the gnats. Swarms of them appeared from nowhere as though they too had seeped up out of the rocky desert with the emergence of day. Their swirling about our faces was no more discomforting than the sharp rocks which jabbed our feet, or the cholla which prodded our legs. But facial oil is less tasty than oil which oozes from wool. The gnats belabored the sheep.

Rosalio called back to tell me that this coming of the "mosquits," as he termed them, would soon give trouble. His prophecy materialized even more quickly than that of the coyotes for, as he told me this, the sheep began slowing to a stop — 2,200 of them — huddling in a valley to bury their faces in each other's flanks and thus outwit the insects.

Rosalio shouted, he tossed handfuls of sticks and loose gravel above the herd, he threw his blue jacket high into the air to frighten the sheep into motion, each time catching it skillfully on his wooden crook. He signaled to the small dog, Boots, which barked sharply, adding her element of fright to a logjammed mass of gnat-pestered wool. Fifteen minutes of maneuvering put the herd into motion but, in one close-packed body, it drifted over a rock-festered hill into the next valley — and clogged again. Rosalio labored once more

We were entering the most gorgeous — or the most awful — cactus country of the entire trek. In some few respects the cactus is beneficial to outside desert invaders, such as the sheep. These animals must be grazing almost constantly as they travel. If ground forage is lacking — and it is just that, much of the way — they nip the leaves from such bushes as cat's-claw or similar plants. And under duress they will nibble the buds and blossoms of the cactus, many times with painful results to their noses. Sometimes a ewe would greedily shred and devour the fleshy portions of a prickly pear.

But certain types of the cactus, particularly the cholla, are bitter enemies of the sheep. As the herd moved slowly on, following their torment by the gnats, they came to the top of a ridge. Ahead of us a gently sloping hill was glowing with opulent but wicked glee — a miniature forest of chollas. The fat individual burrs stood against the sun, back-lighted, and their clusters of delicate long-tapering needles snared the early morning rays, refracting them and holding them, until each plump burr of the cactus — the whole cholla plant in fact — was clothed in a jacket of light. The same effect against the evening sky, five minutes before sunset, turns this jacket of light into a cloth of gold, so that if a poet saw it (a poet new to the desert) he would stir his pen in the fire of its beauty and make an ode to the cholla. It is the cactus with a halo.

It would be mixing metaphors perhaps (and confusing cactus with the herd) to say that the cholla is a wolf in sheep's clothing. Such a statement, too, might be a slander on the wolf. For neither wolf nor coyote actually render as much lingering pain to the sheep on the Arizona desert as does the cholla.

Its serene appearing burrs claw like a cluster of hot fishhooks into anything which touches them even faintly. But even more effectively to ensnare its victims — and for this reason it is called the "jumping" cactus — the cholla sheds its burrs on the ground, littering an area several feet around the mother plant. These burrs cling to the feet of the sheep or are kicked by the herders' shoes up to other parts of their bodies, with every appearance of having "jumped."

Cholla spines, once attached, are difficult to remove. Pliers are the best bet although time, which festers each wound and softens the cactus points, is often the ultimate remedy. Cholla-land is a soft-beckoning hell.

Half an hour after Rosalio had gotten the sheep in steady motion, following their buffeting with the gnats, we came to a promontory which looked out onto a ridge ahead. Thus far the cholla had been isolated but this ridge, widespread and impossible to avoid, was a thick mass of glistening torture.

The faces of the sheep were to the ground, grazing the Indian wheat, nipping fragments of shrubs, as they progressed. The entire herd traveled as a massed unit, a protection against the gnats. The twin enemies of hunger and insects were allied with the cholla in creating the battle strategy for torment. The desert, protective of its silent domains, had maneuvered its forces well in repelling these foreign invaders. One sheep alone would have been hard put to run that barricade ahead. Twenty-two hundred of them churned back and forth in a murderous motion. The herd instinct which pulls the animals close together for protection against any danger was this time playing them false. In massed formation they headed up the ridge.

The instant of an animal's impact with a cholla was marked by a quick stirring and churning motion out of the woolly mass. There, near the front, came a scene of painful drama. A sheep, crowded by others, filled its face with the burrs. The creature jumped upward into the air, head and front feet foremost, above the other animals. Landing back on its hind legs, wedged tightly between other sheep, it spun in frantic jumping circles. Then it dashed forward, partly running on the ground, partly clambering on the backs of the herd. Its agonies were swallowed up in the mass of dumb frightened beasts plunging and flowing by on every side.

Moments later another animal plastered its hind legs and tender rear quarters with a mass of burrs. It kicked backward madly into the air, then with its head toward the ground it was carried forward in its kicking agonies by the ceaseless river of wool.

Above those backs was the roar of battle — the ringing "baas" of pain, the plaints of the distressed, the pitiful "ma-a-as" of the few lambs in the herd. It swelled the air — a cry to heaven.

Rosalio could do little to help. I saw him once in a frantic move for which I only learned the cause when the battle ceased. He ran into the herd, parting the animals as he went, and with his heavy staff, in several careful strokes, struck at a sheep. This happened on the far side and it was hard for me to see. Afterward he told me that the animal had been pushed or had fallen directly into a bed of cholla. Too pained and stunned to move, instead of kicking and fighting its way out as the other sheep had done, it lay there paralyzed. Striking it was the only way that Rosalio had of getting the animal to its feet and headed on.

As this progressed I had a personal taste of what the sheep were suffering. Careful as I was, suddenly as if from the air, two cholla burrs hit me simultaneously in the back of the legs. I tried to stoop for some sticks which would aid me in removing them. Each needle was like a knife blade — being twisted as I moved — which seemed almost dipped in fire to intensify the pain. The sticks were of no avail. Finally grasping my heavy denim jeans with both hands, I jerked much of the cholla free. Some of the spines remained. The pain kept shouting to my senses of the more penetrating wounds, in vital parts of the body, which the sheep were receiving before me.

During this battle in the cholla the animals broke their formation and, as the hill was passed, they streamed in lanes. Probably 200 or more sheep came off that hill with chollas knifing into their noses, feet, stomachs, teats, and backs.

Just one good thing may be said about it all; the cholla is the most painful hazard of the desert but not the most deadly or dangerous. Within a few days after a herd leaves the cholla country, the wounds have all festered, the last remaining spines have come out, and few permanent injuries remain. But while the struggle was in progress it was hideous. The cholla are the guards of this desert, calling "stop" to the invader. The cholla is a delicately sculptured tombstone of Hades.

There are other guards — the rocks.

In a glacial country rocks are often smooth. Or rounded and worn. But no lake bed ever flushed this land. No river or constantly running water has worn the edges of these rocks. They are sharp, broken, rough.

My shoes were heavy, and made for tough hiking; they were miner's shoes in fact. But immediately after leaving the sheep bridge I began to feel the prod of the sharp stones. Before the day was over, the soles of my shoes had begun to show the first signs of wear. What I did not know was that the rocks would not continue this severe for more than a few days. But now I understood why Rosalio had put in not only two pairs of shoes, but a shoe last and soles as well. I would need his services.

I looked ahead at the coming canyons and mountains of this land called Arizona. Dinosaurs and other land giants used to graze on the lush hills that make this state. What the hills were at that time I do not know; perhaps green and moist, perhaps genuinely friendly. But earthquakes broke the rocks, droughts laid low the trees, fires seared the valleys and the grasses, time squeezed the moisture and the ooze away. If one were writing in the language of symbols, as some poets reconstruct nature, one would list these giant saguaros as vegetated reincarnations of the dinosaurs that could not survive. The saber-toothed tigers which once haunted these hills would be the chollas which so gently conceal their teeth and daggers today. The wild beasts of a nameless age have become the wild rocks of a nameless desert. The saguaros and the chollas and the rocks are the only reminders of animate giants of those other times. The greatest beasts of earth could not survive this land. Today the tiny lizard, harmless as the broken stick which it resembles, has taken their place. The meek shall inherit the earth.

At length the sheep, plagued by gnats, lamed by chollas and

boulders, refused to move. Pablo, having packed the camp following breakfast, came up with his burros and met Rosalio in a rounded grassy valley. The herder motioned instructions where to pitch noon camp ahead, which would be our "shading up" place for the heat of the day. It seemed endless time since morning. I looked at my watch. It was 9:00 A.M.

CHAPTER 6

Longest Night

At 4:00 P.M., following the long midday stop, Rosalio started the sheep, which had to be yelled at lustily before they would move.

The movement of the herd along the trail follows a general pattern, which is upset constantly, however, by varying conditions of weather, feed, and water, and such unusual aspects as insects.

In hot weather the sheep, instinctively and without urging, are up and on the march an hour before sunrise — just as the gray dimness of early day seeps across the hills. Rising hour for the herders is between 4:15 and 5:30. At 9:00, on the desert, it has become hot, and by 10:00, often unbearably so. Sometime between those hours the herd begins to slow down. At the latter hour they have stopped — "shaded up" is the term, though it is a concession to the desert to call it so, where the thin web of an ocotillo or the pencil-line cast by a saguaro is usually the only shade. In hot weather it is 4:00 in the afternoon, sometimes later, before the animals will move again, and then they continue normally until 6:00 or 7:00 or 8:00, depending on circumstances of every sort. There are times when they go far into the night. This is the general pattern — resting in the heat of the day, traveling early and late — although it varies more often than it conforms, until on some days there is scarcely a pattern at all.

Rosalio follows the sheep but Pablo, the campero, must adhere to a different schedule. When the herd has departed in the gray of the morning, he washes the dishes, scrapes the leftover food to Boots, and stores the paraphernalia of cooking and camp into the pack boxes.

Then he goes to hunt for his horse and the burros, which may not have wandered more than a short distance from camp but which, even though hobbled, *might* be a mile away. If it is a quarter of a mile he can distinguish their bells, if farther, he often must track them as an Indian tracks a creature of the wilds. The burros retrieved, packing itself is half an hour of grueling labor. Then, marshaling his pack train, he goes to overtake Rosalio.

This morning meeting of herder and campero is an important one, for there the strategy of the day is planned. The campero must go on ahead to set up noon camp, prepare dinner (lunch to city folks), make bread, and do the cooking for the day. But he must locate that camp at a place which the sheep can conveniently reach by "shading up" time. At this morning meeting, Rosalio gives directions where the spot shall be.

The same procedure is followed at night, Rosalio leaving at 4:00 P.M. with the sheep, Pablo packing and following afterward, then meeting the herder somewhat later to get final instructions as to where the night camp will be placed. Here again Pablo goes on ahead to build the fire, unpack, and prepare supper.

After any difficult passage through areas where animals may have become lost, Rosalio makes an informal check of the herd by counting his markers — the black sheep, goats, some of the bell sheep, and others which he can easily identify. The few goats are taken along to act as leaders in difficult areas. The presence of all the markers indicates the probability that there has not been a loss, or cut. If even one marker is missing, a search is started at once.

After passage through the most severe terrain, a formal count is necessary, which can best be made in a dry riverbed, a barren canyon bottom, or similar natural passageway. After the sheep are bunched together, Rosalio and Pablo stand to form a gate, at the same time rousing the herd by tossing stones, so they will start running between the two men.

Rosalio does the actual counting, by twos, yelling out "cincuenta" when fifty pairs have passed, and Pablo registers each hundred by picking up a pebble or notching a stick with his knife. The herder's staff which Rosalio carries — and which he loans to Pablo during the counts — is filled from crook to tip with notches.

On this afternoon of April 20, the sheep started, though with much urging, at 4:00 P.M. Chollas still infested the day, one canyon being nearly as bad a battleground as the cholla hill of the morning. The cactus are thickest on this first section of the trail, the first few days out from the sheep bridge.

The gnats, in the scheme of the herd's movement, were worse than the cactus. Normally, Rosalio told me, insects are a slight problem on the trail but this was a bad year; it had brought them out. Rosalio pointed out a ridge a mile or two ahead, somewhere beyond which Pablo had already gone to establish night camp. The terrain was a jigsaw puzzle of ridges, dry washes, valleys and rounded hills — all strewn with cactus and cat's-claw and palo verde — so I got careful directions as to where the camp could be found. Then, taking up a trail in a dry wash, I bypassed the herd and headed out.

Walking was hard. In the wash the gravel slid with each step, like walking on roughened marbles, and the cat's-claw hugged the way and raked my clothes. On the ridges the sharp rocks and the cholla threw me back. By the time I had reached the ridge of Rosalio's designation the sun had turned in its time for the day. My feet burned; my shoulders ached from the weight of the gear I was carrying. Climbing upward to spot Pablo's campfire, I let the glory of the sunset soothe my wounds as I waited the darkness which would make the fire easier to observe.

But no campfire was visible. There in the wash some distance back was the faint dust cloud of the herd but as I waited it seemed to come no closer. Nothing but semidarkness lay ahead.

It would soon be too dark for safe walking and on the ridge where I stood, with the blackness slipping in like a sea washing up, there were a thousand cholla. The suffering which they had caused the sheep had built up in me a fear of them. Rather than risk the uncertainty of hunting ahead for Pablo, in a blackness bedeviled with cholla, I beat a retreat back along the wash, slipping on the rocks and scratching myself on the cat's-claw, until I was with the sheep once more. It was just after dark when I called to Rosalio and he heard me.

The gnats were driving the sheep in mad circles; their progress was slow. Counting on that precious hour of travel just before sunset, Rosalio had ordered the camp too far ahead. Yet the distance had to be accomplished, for Sycamore Creek — the next

41

water — lay far distant and it had to be reached before the herd was frantic with thirst.

Rosalio and I together wrung a bit of motion from the herd. Then, with full darkness definitely established, the animals began to travel better. The herder came up to me.

"I wish you with Pablo," he said, in his broken English. "Once, when I work for Mr. Kister, I no get to camp on a time like this and had to sleep all night with the sheepsies."

I assured him against worrying about me. Coming again to the cholla ridge which I had climbed at sunset, we started another ascent in hopes of spotting camp.

I would almost as soon have scaled a hill littered with hot coals. Ordinarily the stars would have candled our path to some extent but an overcast sky dimmed even the brightest planets. For a while we worked around the cholla but the suspense of each step was nearly as cruel as the bite. Then Rosalio got a cholla into his hand, which pained severely. And, though less painful, I was struck from behind by one in my leg.

We touched fire to a dried cholla. It seized the flame and, feeding it with a hidden grease, wrapped itself instantly in an ascending spiral of fire. The brilliance from that one guided us easily for thirty or more feet, where we ignited another. From the top we failed to see our campfire, but we made a flaming path up the hill, then down again at right angles. Looking back at it from the wash below, our course was like a candle path to some shrine on the hilltop. We had made a pilgrimage along a path of flame.

But we hadn't spotted camp. It was decided that Rosalio should stay with the sheep — all night if need be — and I would go ahead (this time with a fresh set of directions) in search of Pablo. Striking a match — for down in the wash there were no more cholla to burn — Rosalio, with his staff, drew a crude map of the area in the sand of the dry water bed. But the sketch, designating some branch washes here and some ridges another place, was nearly as vague to me as hieroglyphics and I realized how inadequate for my purposes was the English with which the directions were given. Before starting me off, Rosalio lit some matches and stooped to examine various scuffings in the trail.

"Here," he said at last, "the tracks of Pablo's burros and horse. Fresh. Follow these."

There were marks of burros and horses and sheep of other outfits along with those to which he pointed — all in a faintly visible confusion. So he added: "Pablo's horse is the one with shoesies on. You follow that." Even in the sober circumstances, I burst into laughter. The tracks were plain to the herder but, even in the daylight, I had scarcely been able to distinguish between the marks left by Pablo and those of other camperos. Now at night . . . with matches. . .

But, I started out.

Little gullies etched out by water of flash floods and bordered with desert growth traced their way here and there. At any single time, several such paths were open to me. The wash branched and other washes led in. The tracks of some burros went one way and some another. I didn't bother striking matches to locate the marks of the "horse with shoesies."

The greasewood, or creosote bush, stretched its slender fingers across my way in the darkness and slapped me as I passed. The cat's-claw shrubs, living up to their names, gleefully dug their fingernails into my shirt and jeans.

It was impossible to see where I stepped, and rocks punched a nail in one of my shoes up into my foot. I was progressing poorly, and debated the idea of sleeping where I was without food or cover — casting in with the lizards for the night, as it were — when a light danced and bobbed into view ahead. It was Pablo with two flashlights.

Camp was still more than a mile away. While I squatted on the desert floor, Pablo made his way back to Rosalio, leaving him a flashlight, and returned to me in half an hour. The two of us then made it into camp but, even with a light to guide us, it was a wicked trail at night.

I refreshed myself with two tomato cans full of water and ate a supper which had now cooled for the second time, then tumbled into bed with all my clothes still on. Cactus spines and prickers were penetrating everywhere — in my bedroll, blankets, clothes. In the morning they would be decorating our frijoles and floating in our coffee.

Rosalio stumbled in at 10:15 P.M., having left the sheep still some distance back. He was too tired to eat but he drank water until he was nearly ill. For a few minutes before spreading out his bedroll he sat by the fire, head in his hands, and coughed a number of times. In all my experiences with him I had never heard him cough this way before. Then he slept.

An hour later there came a disturbance which wakened us all. Through the day the sheep would not travel because of the gnats; now at night, urged on by hunger and thirst, they would not stop. The whole herd, blatting and crying, their bells jangling, had kept moving after Rosalio left them, and now swarmed like an army of giant ants in and through our camp.

Rosalio was up and dressed by the time I was awake. Grabbing the flashlight, he rushed out into the darkness ahead of the sheep. He spent half an hour out there in the night, rounding them up, getting them stopped and bedded down at last. It was midnight when he sought rest in his bed once more. Barring further trouble, he would be free to sleep until 4:00 or 5:00 in the morning.

As I was drifting back into slumber I thought about the nearly twenty hours since we had started the day — hours beset with cholla, rocks, gnats, confused dry washes cloaked in darkness, and a dozen other hazards. "How does Rosalio do it?" I mused groggily, then dropped into sleep.

Rosalio, shepherd of the herd.

Upper—Lush Salt River Valley is winter home for the sheep.
Lower—Shepherded by Rosalio, the herd heads out from pasture near Chandler, Arizona.

This little fellow is left behind. The lambs, born in spring, do not make the trek.

Upper—Crossing of the dangerous Salt River is made on a suspension bridge built for the sheep.
Lower—The pack burros take their time in crossing.

This sheep makes a giant leap in its relief at leaving the swaying bridge.

Upper—First part of trek is in Tonto National Forest but here it is a forest more of cactus than trees.
Lower—Rosalio removes a cactus spine from the paw of his dog.

Upper—The sheep huddle together as a protection from gnats, making it hard for Rosalio to keep the herd moving. Lower—Midday camp is a time for cooking, relaxation, and chores.

Pablo's tasks include packing burros twice daily, making camp, bringing water, and cooking.

CHAPTER 7

Desert Magic

We were up early, had breakfast of fried eggs, mush and coffee and were on our way just as the sun was rising. Rosalio and I followed the sheep. Pablo stayed to pack camp but discovered that his horse, even though hobbled, had strayed off during the night, going back toward the sheep bridge. By skillful tracking he retrieved it.

During the morning the sheep followed along dry washes, over open cactus and greasewood country, and both along and across rugged canyons. Narrow canyons are worse than wide ones, for the sheep more quickly spill out into adjoining country. Once during the morning the herd was following along three parallel canyons at once, and Rosalio had to patrol them all. Boots was of little help here; this was too rough and cruel a country for dogs.

Not sharing Rosalio's responsibility and worries for the sheep, I went slowly, lingering to dwell luxuriously on each aspect of the country.

Six buxom quail came strutting down a filareed hillside, their bay windows pouching and their saucy black faces bobbing up and down like a troop of entertainers taking the stage for a session of songs and jokes. Each had a jaunty, plumed minstrel hat. And behind one of them came the featured attraction of the show — a family of baby quail, each no larger than a generous puffball. Sighting me, all the performers but the mother and her brood made a whirring retreat from the stage, taking to the wings as it were. Assuming instant command of the entire act, the mother convoyed her youngsters along the rotten carcass of a fallen saguaro and

53

then, without a stop or hesitation, ran on up a gentle slope and started calling me names.

Now the chicks were gone. This was no stage comedy; this was the act of a magician. The mother had said "Presto, be gone" and even without the wave of a wand or a wing, the thing was accomplished. I poked with the butt of my tripod about the rotten roots of the cactus. Finally a tight-drawn ball of fuzz, with a shiny pin for an eye, peered up at me; that was all I found. I took a turn to the other side, planning to approach the baby quail from a fresh and more suitable angle, but in that instant it was gone. Another magician. No amount of further prodding turned up any of the brood. Looking now to the mother I found that she too had disappeared. She had been willing to expose herself as long as it would aid her purpose but, seeing that I was bested, she at once took cover, to be ready for whatever next skirmish with danger for her family might come this way.

While I was still admiring the mother's skill and the supreme discipline of the babies, I suddenly became proud that I had been responsible for that one added lesson in concealment. For as I climbed out of the wash I saw, sailing right above it, winging directly over the rotten saguaro, a red-tailed hawk. No doubt Mrs. Quail and her youngsters had continued their morning airing but I smiled and felt contented — as the hawk hovered like a glider — at the assurance they could play their tricks on it as well as me. The hungry bird would have to ease its appetite on less dainty morsels, on life less skilled in magic. Mrs. Quail is a Lady of Legerdemain.

This area, it soon became apparent, was struggling furiously to get me lost and separated from the herd, not so much because of the confused and threaded intricacy of the meandering washes, but because of the natural distractions to progress which it afforded along the way. No sooner had the hawk disappeared in hungry disappointment than I was captured by the wink and blush of a cactus bloom. Twin blooms in fact. Cactus flowers along this trail during the last two days have been almost as thick as dandelions in springtime. Scores of cactus varieties grow in Arizona — a large percentage of all the varieties found in the world — and many of them have blooms in the spring.

But these two that caught my attention were special. They were alone, and their purpled splendor stood out in the grayness of the wash like a fresh-bloused girl in a dingy bus station. Each petal was flawless, unscarred, and waxed with a subtle polish. These blooms were as delicate in coloring, and more delicate to the sight, than a party orchid, but I touched the petals and found them sturdier than the orchid — silk-textured yet firm.

Progressing along the wash, I became conscious of the structure of other desert growth about me. Intricacy was everywhere. Each of the giant saguaros had a thousand hand-tooled needles. Perfection in pinpoint. Every palo verde tree — and they were thick in this part — had a million tiny leaves, each no larger than a baby's fingernail, yet more wonderful to look at than the giant leaf of a palm, just as a Swiss watch is a thing of greater perfection and delicacy than the works of an eight-day alarm. Every leaf of the greasewood, the cat's-claw, the mesquite, was seventeen-jeweled in its intricate perfection.

The wash became a braid of narrow gravelly runlets fringed with bushes, the braid became a skein, and I emerged at last onto a broad dry riverbed — the lower end of waterless Sugar Loaf Creek — beside which the sheep were shaded for noon.

Rosalio and Pablo had already made camp and the red kyack boxes stood in neat order, with the fiery water kegs at the end, beneath a palo verde tree bursting with golden blooms. Each ordinary act of the noon camp — mixing pan bread, washing clothes after dinner, shoeing the horse — was a drama in a majestic setting of color. We were shaded up in a sea of saffron splendor.

CHAPTER 8

Stars, Solitude, Rx

If I made notes now and wrote a book upon returning home and to civilization, it would be done largely by the sweat of the forehead. By writing large parts of it here, on the trail, much goes down without effort, my pencil needing speed just to maintain the flow. The pointed lead moves almost as though some Ouija of the silent sands were commanding it. Concepts come, stirring up from the earth, coursing from the skies, flying in with the wings of birds. I am merely the screen — the web — which snares them, condensing them in script.

It is night. Only a few glowing embers remain of the evening campfire. Rosalio and Pablo are both asleep. I was asleep too, but something wakened me. Drowsily I looked upward, then — unbelieving — rubbed my eyes. The whole sky was a carnival of dancing lights.

Not enough has been written about Arizona nights. In his book, *Midnight on the Desert,* the English author Priestley observes that "Arizona is geology by day and astronomy by night."

It is understandable that Priestley wrote his most philosophical book under a nocturnal spell in Arizona. As I looked overhead the skies blazed and burned. It was a fiery, glistening, wild heaven. The whole earth — and I an attached part of it — was bathing in the starlight, and something inside me seemed to be expanding like opening petals — a cereus which blooms only at night.

I could not sleep. Whole new worlds swarmed around me. The hard desert where I lay seemed magnetized with strength. The air was gently warm and I threw back the covers, opening my bed to the breezes blowing up from the Salt River far below. By now, Jupiter was a spotlight, streaming its blaze earthward. I got out my notebook and in the brightness of the stars — stars which I had really not seen for years — I began writing the secrets of the night about me.

During the first part of the trek, Jupiter became a new and silent friend. Glowing like one of the coals of our smouldering campfire, each night it would wheel into the sky from the southeast at 10:00 P.M., followed closely by Scorpio, then the Milky Way.

The Milky Way. When I had seen it first, toward the very beginning of our journey, I had mistaken it for a long soft cloud. The next night I purposely wakened after midnight. By the light of the stars alone, the tiny face of my wristwatch was clearly visible. But no watch was needed. The Big Dipper, brilliant timepiece of the sky, pivoted overhead about the Pole Star. Never was I more than ten minutes off in my judgment of the time as I read its star-studded dial. The sky was hot with points of light. They didn't twinkle, for the night was clear and a breeze had swept away all trace of haze and dust. They burned steady and bright.

There was a cloud in the east. But the cloud kept its shape. It was self-luminous. The delicate stems of an ocotillo near our camp swayed against the brilliance of this cloudlike formation. Only slowly did I begin to realize that it was not a cloud, but a galaxy of distant mysteries.

Stars I can understand. I lie awake in awe beneath them. But the Milky Way is almost too much; it is more than I can even hope to grasp or comprehend. A star is a definite point of light; the Milky Way is a hazy symbol of inscrutability. Thought expands to the infinite in its very contemplation.

I spoke of Jupiter as a silent friend. Being a planet, it may indeed be silent. But the stars which surround it are not. Lying on my back on a moonless night under Arizona skies, I had a sensation of music from the sky. I analyzed it thus.

After having separated each sound of the night which I knew — and classified it as best I could into bird call, or wind, or breath of trees — there was still a constant undertone of singing vibration, low but distinct, which filled the air.

Perhaps it was an accumulation of tiny unnamed noises — the breath of a million insects, the disintegration of a million rocks, the tag ends of a million pulses from the other side of the earth. But, that undertone of constant singing was there. Absolute stillness in the world, I discovered, did not exist. That singing of the night — for such it was — which I heard along the sheep trail, in my mind easily became associated with the stars, which were a luminous reality at midnight, much greater to the eye and the consciousness than insects or rocks or the far side of an unseen earth. (I have since learned that definite sound vibrations can be picked up from the stars by delicate instruments.)

During these desert nights — although it was equally true by day — came also my first comprehension of the values inherent in solitude. To be torn for a time from all contacts with civilization can in itself be a civilizing experience. Here on the sheep trail is one of the most abundant of ingredients — yet one of the hardest to capture in normal life — simplicity.

No newspapers, no headlines, no telephones or radio, no noise of traffic. Until one escapes these things completely it is hard to know their debilitating effect on original thought.

Some of humanity's greatest spiritual discoveries have been made by tenders of sheep or dwellers in wilderness. Most great seekers after truth have sought solitude. It is scarcely an accident that some of the greatest religious concepts were born, in solitude, under open skies. Moses was at one time a herder of sheep; he gave us the Ten Commandments. Amos also was a herder of sheep; from him came the first concepts of a God of justice. Mohammed tended sheep and camels in the region around Mecca. Buddha, in his search for truth, lived as a hermit. David, giver of the psalms, was once a shepherd, and so was Elijah. Jesus went into the wilderness and for forty days was tempted, being offered all the kingdoms of the earth to test his purposes. Those forty wilderness days — and nights — were the womb for concepts which today are heralded as tenets of the Christian faith. No one with his powers and relations to God could spend forty days with the solitude and stars of the desert without receiving such a glow of new life as would change the world.

The symbol for the planet Jupiter, given it by desert dwellers before the pyramids were built, is Rx. That symbol is today a sign of healing. Stars and solitude are among the healing treasures to be discovered along the sheep trail in Arizona.

CHAPTER 9

Morning Star

If Pablo did not call me each morning I would sleep right through breakfast. But he has acquired a habit of waking an hour before daylight and as soon as the fire blazes and the mush bubbles in the Dutch oven, he stirs camp with the shout: "Fran-cees, Fran-cees." It came at 4:15 this morning.

Morning — even in the blackness of night which constitutes the sheepherder's rising hour — is not my usual time to view the stars; the stir of breakfast preparations and the brilliance of the campfire rule against it. Jupiter and Scorpio, when we arise, are still blazing in the southwest, but the time for viewing them is in late evening or the dead of night, when all is still.

But this morning Rosalio called my attention to a planet which we had not seen before, just rising beyond Four Peaks in the east. It was Venus.

"Lucero," said Rosalio. "In English you call it Morning Star. In Spanish we call it Lucero. We had a song about it."

In his rich gentle tones he proceeded to sing:

Lucero de la mañana.
No te des a conocer,
Que por ahi andan diciendo
Que reinas mi porvenir."

Then I realized that Rosalio's last name, Lucero, was the Spanish word for Morning Star. "Many our people named after stars," he explained, and he spoke the name of Juan Estrella, another herder, as an example.

The sheep were bedded on a southwest sloping high hill a

61

quarter of a mile from camp. Breakfast was hurried and as the twilight of morning pushed in upon us from beyond the mountains, Rosalio picked up his staff. "I go back to sheepsies now. They pretty good last night. I only had to get up once."

His words sounded precisely like those of a caring parent concerned over a colicky child. Rosalio is not only a shepherd, but mother and father to his 2,200 children.

I saw his dark figure dimly working among the herd. Jupiter and Venus were the only lights left in the sky, which was paling fast — the evening and the morning stars. I looked at Venus working its way above Four Peaks; I looked at Rosalio working with the herd on the southwest slope, then broke out into sharp gooseflesh, which always happens to me when I am struck with some deep emotion. Rosalio Lucero; Rosalio, the Morning Star. Here was a name that fit. He was one of that chain of desert dwellers nurtured by starlight that had enriched humanity since time began.

As soon as Rosalio left for the sheep, Pablo went off in search of his horse and burros, leaving me to guard the camp in case the burros beat him back. Pablo assures me that his pack animals will eat an entire camp if given the chance. I have in the last day seen them eating rope; chewing and swallowing newspapers, printed in either Spanish or English, it makes no difference; and at least having a try at tin cans. Pablo, with droll humor, claims they will only eat the labels of cans which have the pictures of food — peas or tomatoes or corn.

Here alone in camp, sitting by the kyack table, still half an hour before sunrise, I can see the herd coming. Rosalio has gotten the sheep in motion. Their bells jangle as the gray backs wend in a hundred paths through the brush and shrubs down the low mountainside. A canyon intervenes before our camp. The leaders are already disappearing into it — a rocky steep-walled canyon. I know, for I climbed it when coming in last evening. Lucky the sheep were not too restless in the night or Rosalio would have had to climb that canyon repeatedly in the darkness. Now the animals will test the strain of their wilderness ancestors in accomplishing it.

Burro bells — more metallic and louder — are coming in now from the east, off toward Four Peaks. Pablo has traced his pack animals. I have been writing by the light of the campfire, though now at 5:40 A.M. the sky in the east has acquired an ample glow.

This I will remember as the night of the gentle rain. The sky was woolly with clouds as we had gone to sleep. At 10:00 P.M. by the Great Dipper I'd wakened with cold raindrops spitting in my face. Quickly rustling my cameras under blankets, I likewise pulled my head under the tarp, which was stretched tight, making the rain sound like drops on a tin roof.

This was but a shower and as the first bout ceased I came from cover to observe what took place. A groundhog poking my head from the burrow to scan the weather. Those things which come most frequently are perhaps least observed. In all the storms which have wet me or the roof above me I have never before taken occasion to observe one as a *show*. This one I did.

We were camped in a desert clearing cupped by hills close by, with mountains forming a second low circle beyond, like a saucer set in a plate. Above were bright stars but over the western hills a darkened mass appeared, dark clouds shaped like a hand — a witch's hand, it seemed. The long black fingers shoved out from the black hills, erasing one star after another. A witch was catching wayward children. Soon the points of light overhead were rubbed out and cold pellets of rain began falling. Then slowly the fingers vanished and the starry-eyed children played again.

Scarcely had Jupiter moved the distance between two fragile branches of an ocotillo which stood between my view of it, than a ball-shaped mass of dim blackness — another rain cloud — forayed in from the southwest. It crept along the south horizon, blurred Jupiter itself, then screened it out. There was rain from that cloud but not on us. A regiment of witches groped along the eastern hills but no move was made to ride toward our camp and finally the witch dance was over and the sky came clear again. Just one flash of lightning had been discharged.

Over on the high slope, Rosalio — the Morning Star — was heading his sheep out to start the day.

CHAPTER 10

Water, Salt, and Fences

Rosalio and I made our way together with the sheep. He told me that Sycamore Creek was below somewhere and I was anxious as soon as possible to go ahead and choose a grandstand seat to watch the animals head down for water.

Approaching this watering place, a mountain slope which constitutes the trail starts downward at a grade which is difficult to hold to, from an elevation high above the stream. The descent is cut with ravines and strewn with loose shale, and grown over with saguaro, ocotillo, and prickly pear. Half way down I chose a jutting rock and sat to wait the herd.

They broke over a squared hill, slowly, and began pouring down like grayed molasses burping from a pitcher. Fifty molasses-like streams descended in different courses, twisting, stopping, now bursting ahead in gushets of speed.

Their final pace was frantic as their thirst — built up from hard days of desert travel without water — propelled them toward this creek.

As the mass of sheep plunged downward, I realized that these animals were borrowing the instincts of their wild mountain ancestors. Surefooted, these modern wool bearers leaped their way down a rocky slope to water. As they reached it, prolonged and repeated baas of thankfulness and contentment came from each animal. Their call reached up to me on my grandstand seat and soon the air was filled with their simple expressions of joy.

Sugar Loaf Peak rises above Sycamore. This area is one of the high points on the southern part of the sheep trail. For Rosalio

65

and Pablo, for the sheep, and for me also, it was a figurative high point as well.

Gunnar Thude's foreman had come up the Bush Highway by truck, then bumped in on a torturous road to meet us, bringing camp supplies to replenish our dwindling provisions. Pablo was obviously pleased. "If you didn't come," he said jestingly to the foreman, "Fran-cees would soon have to be eating rattlesnakes."

Occasionally, the owners fail to make connections with the herds at the few meeting places en route, and the herders are forced onto short rations.

A letter had also been brought in from Rosalio's sister in Albuquerque, and some Spanish newspapers, which Rosalio and Pablo read eagerly. The papers had been published in Santa Fe many days before, but the contents were news to them.

This bivouac at Sycamore was a double blessing for the sheep. Not only did they get water, but they got salt as well. The foreman had brought in sacks of rock salt, which Pablo and Rosalio scattered out, as the animals went for it almost frantically.

Rosalio had even greater occasion than Pablo for elation. One of Gunnar's other herds of 900 sheep in charge of Manuel had come in from Cave Creek on the Bloody Basin Trail and had been waiting here for four days. In a giant maneuver in which all of us — a total of six men — participated, we cut some 700 animals from Rosalio's herd and joined them with the band in charge of Manuel. A new count showed that Rosalio would now have 1,547 sheep, a more manageable number. "Better for me," he mused, "also better for the sheep. If not much feed, more can eat."

From now on, Gunnar would have three herds heading along the Heber-Reno Trail. A caporal, or captain, on horseback, would be brought in to supervise the three herds as, traveling several miles apart, they headed farther and farther into the wilderness.

When our sheep at length reached Bush Highway this wilderness, stretching before and behind us, had a brief respite.

Varied signs of civilization, from which I had been separated for a week, were crowding in on all sides. A car went by on the bumpy road, the first I had seen since the sheep bridge. With the foreman, I rode up to the Sunflower General Store, where I met the first person, other than the herders, since leaving the bridge. And as we and the herd left the road, heading on into the hills once more, we followed along a three-strand fence — the first fence of any kind which we had encountered. But this, Pablo told me, was just a "drift" fence or "wing" — a barrier to keep range cattle from wandering too far to the south, for this sheep trail traverses cattle country in this area. It was not a fence to exclude us; the trail in fact went right through it and the sheep went under, for the bottom strand or wire was purposely not barbed.

As we climbed higher and higher we were able to see the Bush Highway, which extends from Mesa and the Salt River up to Sunflower and Payson, threading through the rocky terrain down below us. During the two hours that it was in our view only two cars went along it. But to me, who had become acclimated to wilderness, they were foreign intrusions.

We plunged into a somewhat rough and rocky section of hills and made camp well before dark on the slope of a broad north and south valley with a high mountain — Boulder Peak — to the east. The highway, cars, and store — even the drift fence — were now lost from view. We were back in the natural world.

As if to prove it, Pablo motioned me to come look at a tree close to camp. Around a hole in the tree's side a small colony of bees was working. Pablo put his hand into the hole; several bees lit on it, without hurting him. "They here every year," he said. After the evening meal, from his new supplies, he put some canned apricots and sugar in the hole. "It keep them busy," he explained. "Otherwise, they might drive the burros from camp."

Just before reaching camp, Rosalio and I had come upon a nearly two foot long Gila Monster, the first we had seen. Rosalio said they were less dangerous than they looked. "They bite. Maybe you get sick. But you don't die," he explained. It is the most colorful member of the lizard family I'd ever seen.

As I cleared a place for my bedroll in camp I found a strange spider, a dull gray and nearly the size of a huge elongated bean. Its back was covered with tiny egg-like warts or bumps. Gently I removed it from the spot which I had chosen for my bed but as I did so, the warts began to move. Looking closer in the dim light, I discovered that they were baby spiders. Dozens of them. The mother had a whole family riding about with her, piggyback.

In the space of a few hours we had exchanged stores and automobiles and fences for Gila monsters, wild bees, and spiders. These were more in keeping with the nature of the trail.

CHAPTER 11

Shepherd of the Trail

Rosalio never carries a gun. Hundreds of bears and other wild animals roam this area. If they threaten his sheep, he gathers the herd closer into the camp at night. He has no fear of wildlife.

In making his bed at night in desert areas he carefully removes scorpions or tarantulas that may be in the way, brushing them gently aside but never killing them.

Rattlesnakes live along the length of the sheep trail by the thousands. If one is found in camp, occasionally Rosalio will kill it, but seldom otherwise. "They no hurt me," he told me, "so I no hurt them." "A rattlesnake is a gentleman," Pablo once said. "He gives a warning."

Rattlers are seldom a problem to the sheep. The normal movement of a herd will trample any snake to death, so they take for cover. One of our lambs was bitten by a rattler. Its nose swelled and turned black. Rosalio tried to catch it to give treatment, but was never able to do so. Eventually the swelling went down and all was well.

Rosalio's unassuming fearlessness and Schweitzer-like regard for all living creatures was, I realized, beginning to rub off on me in various ways, which made this wilderness experience exhilarating to me, rather than fearsome.

My discovery that his last name meant "Morning Star" in Spanish, seemed to strengthen the bond of understanding and friendship between us. As we started our morning progress over the calloused hills together he reached farther and farther back into the springtime of his life and etched its pattern for me.

69

He told of his grandfather who had helped drive cattle along the old Chisholm Trail through the wastes of Texas, up to railhead in Kansas. I remembered the Kansas state motto: "To the stars through difficulties," and realized that some of Rosalio's heritage was involved with that saying.

He had an endearing way of speaking of his grandparents. "My papa's mamma," he said, or "after my papa's papa drove cattle, he drove oxen carrying freight from Albuquerque to Kansas City."

Rosalio's first experiences with sheep came when he was eleven. His father was a herder near Albuquerque and the boss, taking a liking to the boy, suggested that he go along to help, at ten dollars a month. At fourteen, he herded sheep near Acoma. "Sand knee deep on south side, getting down from Acoma," he told me.

The accidental turning of a saw in a New Mexican mill when he was fifteen, diverted Rosalio permanently to herding. To my questions as to whether he had ever married, his simple reply comes (and the reply is a biography of his life): "I wanted to marry sometime, but when I had sawmill accident, I could no get regular job anymore."

The accident had partially destroyed the use of one hand, and he had turned to the thing he knew best. Steered into his life course by the literal wheel of chance, he started devoting all he had to the thing he was doing.

The sheep trail has drawn men to it for a multitude of reasons. Some come just for the job, but these do not return a second year. Pablo's glorious manifesto that "here at least you work in the sun; in the city you are always in the shade," is for many of them a strong reason. The life on the trail, though hard, is a life that one can love — the everyday pitting of oneself against the harshness and grandeur of nature.

For a campero, the work is stern. For just an average herder in charge of sheep it is even harder. For a deeply conscientious man like Rosalio, the task of shepherding a herd of his "children" over one of the cruelest trails in the land, is a responsibility equal to some of the more critical jobs in American life.

With Rosalio, sheepherding is not really a job. It is craftmanship — an art which has settled almost into an inherited instinct in handling the animals. Though he says no, I have a feeling that

70

Rosalio has some of the blood of the New Mexican Indians fortifying the blood of his more dominant Mexican and Spanish ancestors. I believe that the instinct with which he tends the herd goes back not solely to his Mexican father watching his sheep on the hills surrounding Albuquerque but stems, too, from Indians tending flocks outside the walls of their mud villages in the days which followed the introduction of sheep from Spain. Back along separate routes of blood and tradition, Rosalio's craft as a herder extends to the hills of Old Spain. Those Spanish flocks also trekked each spring from lowlands to mountains, along given trails, as in Arizona today. This man's profession is ancient and through the ages has been among the world's important occupations.

Late in the afternoon, Rosalio had an unusually frustrating tussle with the herd. After it was over, he spoke to me slowly, revealing his age as he did so: "Sixty years too old for this trail," he said softly. "I may have to leave the sheepsies some day."

When he quits for good he plans to go back to his New Mexico where he started herding, to retire and rest, and dream about the days and nights and joys and hardships with his sheep.

He told me that once he owned a little house in Old Albuquerque and rented it out. "But I have much trouble. Once the people no pay rent. So I sold the house. Less trouble that way. More better, don't you think?"

Rosalio's watch steadily loses time. This afternoon, shaking it gently, he said: "It goes all right in the daytime, but it sleeps at night."

The episode which caused Rosalio to lament his age had not been so much a test of his physical strength as it had of his inner emotions. Remorse, not exhaustion, had triggered his reaction.

There had been time-consuming difficulty again starting the sheep, when the hour arrived for afternoon departure. Down and around a hill, and up again, Rosalio ran, working a small band into motion, only to have them circle back and stop. Obstinacy seemed to possess them. He called on the aid of Boots and he himself barked like a dog. Nothing availed.

In many countries, on many trails, if a herd does not travel it means inconveniences. On this trail it could be more critical. Sycamore Creek was two days behind; the next water was still more than a day ahead. Even now some of the animals were fretting for it. A schedule of some sort — a water schedule — had to be maintained.

71

Necessity demanded action. Slowly, as though it seeped out of the ground or was transmitted by the herd itself, a flame of temper rose within Rosalio. His face revealed it clearly. Here was a case of real anger. He rushed into the herd and slid his staff over the backs of the animals. Then he struck the most obstinate ones several times. Not alone the blows, but the fire and energy of his rage, transmitted to the sheep and they started. The spark which had set off his anger had been transmitted to him from the herd; now his anger created something of an electric impulse in the air and went back to the animals. It is the only time that I had ever seen Rosalio lose his temper.

The animals moved. In a close mass they crawled into a ravine, on the far side they came out on a score of rocky trails, and fanned out over fifty acres of hillside, working gradually toward the top.

Rosalio — tired, exhausted, but having got the herd into motion — sat down to watch and to rest. I was with him. His dark face had the look of the sea in it — shiny with sweat, and slowly settling back from the darkness of fury to his usual calm.

"I never had such a herd in forty years," he said. "I like my sheep to go slow. It is good for them. But these just make circles."

At the far end of the ravine the sheep were discharging frantic and pitiful cries into the air, filling this whole section of mountain with their bleating.

"What's their trouble now?" I asked, not comprehending.

In his answer I saw that Rosalio was thinking deeply about what had occurred. It was evident that all anger was gone. It hurt him — injured him greatly — to realize that he had lost patience with his "children."

"They baaing for water," he explained. "Water was down there early in spring and maybe the ground still wet. They smell it." Giving a close ear to their pleading, he continued: "I get mad with the sheep. Maybe they get mad with me now that I can't take them to water."

With his red bandana, Rosalio had wiped the wet from his face. His hat was pushed back, revealing his black hair, and letting the mountain air flow across his forehead. Now his brown features were calm; his deep quiet eyes were farseeing and gentle again. Once more he was the shepherd of the trail.

CHAPTER 12

A Poet Is Born

Formerly there was a six day stretch between watering places at Sycamore and at Tonto Creek, and numerous animals perished. So the forest service piped water down into a large storage tank, variously called Rock Tank or Bushnell Tank, after the ranger who supervised its construction.

Back at Sycamore I had been forewarned as to what to expect from the sheep. But here, I was caught by surprise.

As they approached the Tank, the herd plodded along slowly, grazing when any forage was to be had, heading down into the valley. Then an electric spark was discharged. Rosalio explained it to me later. When about half a mile from the Tank, the lead ewes smelled water. They began a mad plunge forward. The impulse carried to the other animals following behind. Soon an immense stampede of wool was streaming down the hill. Nothing could have stopped them. The pattern of fast-flowing movement was weird — hundreds of gray backs streaming and fanning downward. Where rocks or ravines intervened the mass movement was broken, and a dozen separate threads of gray moving life went swirling forward at frantic pace. Here was something elemental, primitive. It was one of the joyous sights of the trip to see them rush up to the drinking troughs and satisfy their thirst.

Our sojourn at the Tank also held surprises and special joy for me. Again the boss had come with additional provisions, bumping in over a jeep trail from a dirt road three miles distant. And with him were my wife Helen, and our ten year old daughter, Adrienne, who had come in to get a brief taste of the trail and to bring me a fresh supply of film. We camped for the night at the Tank, and they had the experience of sleeping by the fire under the stars and partaking of Pablo's fine cooking.

As we ate, I told them of my experiences along the trail. Adrienne was so thrilled that she started composing a poem about it. She reworked it later and the next year, when I had written an article concerning this trek for *National Geographic Magazine* and was going over it with the editor, he insisted on including our daughter's verse. "She has said it all," he observed, "in just ten lines."

SHEEP

Under Arizona's great blue sky
On winding trails the sheep pass by.
Into the valleys, up mountains steep,
Traveling steadily come the sheep.
Down to the river to water they run,
To conquer their thirst from the long dreary sun.
They wind their way through cactus spines,
And journey by forests of tall, stately pines.
In mountain meadows then they roam;
And each sheep has earned its summer home.

I had a day or so of homesickness after Helen and Adrienne, and the boss, returned to civilization. For the herd, the herders, and for me these were days of climbing as we ascended by slow stages to the top of Reno Pass, and to one of the finest campsites of the journey. Ahead and below us, spread out for miles, was Tonto Basin, one of the wild sections of America, with a thin silver thread winding through it, which Rosalio told me was Tonto Creek. Barely visible, far to the right, glimmered the upper end of Theodore Roosevelt Lake.

Usually I followed Rosalio and the herd but on the morning at Reno Pass I lingered behind to watch Pablo break camp.

As he drove his burros in — still hobbled — they took little forward jumps with their front feet, looking almost mechanical — like tightly-wound Christmas toys. "You've got a bunch of jumping jacks," I punned.

Pablo grinned. He has an endearing way with his burros. In the mornings he often asks them if they are mad with him, and rubs their noses and only swears at them gently when they disobey.

He likes to tease them and they seem to enjoy it. Before starting to pack, he took a piece of his breakfast biscuit, held it out for a burro to eat, then quickly snatched it away several times before letting him get it. Gently he rubbed the animal's nose.

Then began the campero's hardest task of the morning. Pablo has skills as cook, baker, butcher, shoe cobbler, and blacksmith, but none of them takes the strength and expertise which is required to pack his burros twice daily. A dozen pack boxes, along with bedrolls, water kegs, shovels, axes, and endless paraphernalia have to be deftly packed onto the seven animals so securely that nothing will shift or come loose, no matter how rugged the terrain.

These packs, few people realize, also carry most of the personal possessions owned in this world by Pablo and Rosalio. Save for one extended vacation each year, they spend nearly their entire lives camping out, on the trail, or in the northern forests or southern pastures.

The packing completed, Pablo fed his horse some oats which likewise had to be carried on the backs of one of the animals, unhobbled his burros, and started out to overtake Rosalio and the herd.

When nearly down the pass Pablo and I saw a lone rider approaching us on horseback. Pablo recognized him, and as the man stopped to greet us, said to me: "This is Marshall Loveless. He can show you some bears. You two ought to get acquainted."

At a later time I did become well acquainted with Marshall Loveless and his brother, who lived in a cabin a mile or so from the trail, near the ruins of old Fort Reno. The Loveless Brothers were big game hunters and fur trappers; it is this latter occupation which brought them to this area of the sheep trail. Marshall told me that years before, he had accompanied Buffalo Jones on his famous hunting trip to Africa and had later guided African hunting safaris on his own.

I had no way of knowing whether his statements were authentic. But several years later, after showing my film of the sheep trek at the Santa Barbara Museum in California, Max Fleischmann, the yeast baron, came up to me, and said: "You showed a picture of

Marshall Loveless. He was my guide on a hunting trip I once made in Africa."

Getting down to Tonto Creek from the lower part of Reno Pass required a hideous descent. Once Rosalio, intent on watching his sheep, fell and cut both of his lips severely. In places, nearing the bottom, the going was so steep that we slipped and slid part of the way down. There was another mad dash for water, even more spectacular than those at Sycamore and Rock Tank. There was another road, but no cars came along it to disturb the silence.

CHAPTER 13

Killing Heat

It was upward from Tonto Creek — up into the foothills and then over the sky-tabled plateaus of the Sierra Ancha range, over the very top, into the remotest section of the trail. It is a nine day journey from the Tonto Basin to the next signs of civilization, in Pleasant Valley.

It was upward, first through mesquite which was close grown and hard to negotiate, then onto grand wide tablelands, floored with dry filaree and haphazardly sprinkled with palo verde and isolated cactus. The next higher level produced greasewood shrubs, large and slender-stemmed, which atomized the air with their pleasant fragrance. Lizards played hide and seek with Rosalio and me as we made our way upward. Grasshoppers measured our strides, jumping six feet ahead as we came along and then waiting, daring us to step out as grandly as they.

There were two dogs with our herd now, Mulatto's "Georges" having come in for a few day's visit. And our own dog Boots was working much better with the sheep. Following Sycamore, Boots had sunk to the nadir of accomplishments as a sheep dog and after a day of uncertainty as to the cause, Rosalio had discovered the reason.

"Boots wants another dog," Rosalio explained. "What you call it in English? She is warm."

Boots was in heat.

And now, leaving Tonto and climbing into the hills, Boots and Georges romped gaily into the ravines and up the slopes with the sheep, doing an adequate job of helping where help was needed.

"Georges catched her yesterday," Rosalio explained seriously. "Now Boots happy. She makes puppies. In about two months, we have some little sheep dogs."

Our calculations indicated that the blessed event would come around July Fourth. We began thinking up some appropriate patriotic names. The dogs worked so well together that we arranged for Georges to stay with us off and on during the rest of the trek.

Making a giant squared mesa our objective for the night, we saw up on its top the first sprinkling of junipers. The altimeter of changing vegetation was registering our ascent.

The night was warm, the day broke hot, and as the sun climbed so did the heat, working us into perspiration.

At noon camp it was searing hot, with only two thin palo verdes, plus a few scattered low shrubs, to give us shade. Burros, dogs, and humans — each of us has sought a patch of it to bar off as much sun as possible. The heat is beating down from above and flowing up from Tonto Basin. A tent would only trap it. There is nothing to do but sweat.

After eating, we lay in the thin patches of shade, having to let either our heads or our feet project into the furnace of sun rays. Down below, the herd is broken up into tiny bands, each crowded about a bush or shrub, seeking shelter as best it can.

My feet have been in the sun; I pull them in and remove the shoes. One could fry an egg — sunnyside up — on the soles. On the stones about camp the egg would be well done.

We have no thermometer save the rising mercury of perspiration, but it is not hard to guess the temperature. In Tonto Basin it had been 110 degrees and this day is hotter.

The degree of heat which the sheep must endure as they graze, noses to the ground, is revealed in a letter I have since received from the director-naturalist of the Living Desert Reserve, Palm Desert, California.

The letter says, in part: ". . .when the air temperature reaches 120° it is quite possible for the ground surface temperature in the sun to reach 180°. When we hear a weather report on the radio it gives the air temperature five feet above the ground in the shade. Obviously, when you are out on the desert hiking there is no simple way to levitate yourself five feet above the ground in a shady area!"

Upper—After three days without a drink, sensing water ahead, the sheep race downward.
Lower—Rosalio and his dog watch as the herd heads toward water.

Upper—At Rock Tank herd owner brings author's family in for overnight visit. Line's daughter offers a salt treat. Lower—Goats lead herd through terrain where sheep refuse to go except as followers.

Upper—The pause that refreshes. Lower—In wild terrain, counts are often made to assure against losses. This sheep likes to get it over with speedily.

Rosalio refills the water kegs at Tonto Creek.

Upper—A Dutch oven placed on coals is used for baking biscuits.
Lower—Hobbled burros hungrily eye meal preparations, hoping Pablo will turn his back.

Upper—A baby burro, born during the trek , is soon able to tag along behind its mother.
Lower—This burro seems utterly dejected at the prospect of another hard day.

CHAPTER 14

Borrego Plunge

In the Mexican settlements of Los Angeles, Phoenix, and other southwest cities this is a fete day — Cinco de Mayo. May 5. Mention was made of it in the conversation at breakfast — the subject was brought up by me, I believe — but here on the trail this was just another day with the sheep.

From the point where Rosalio had left them, the herd had worked nearly up to camp in the night. It takes skill in determining a camp location so that the terrain will naturally lead sheep toward the site rather than away from it, in case they might move in the darkness.

We had spent the night on the summit of a great ridge which from below had looked like the summit of the entire range. But within an hour after starting out in the morning, that ridge and our campsite were far below us. We were now following along the red rock gorge of Canyon de Borrego — a magnificent blood-dyed scar in the green-clothed mountain. Far across the canyon, three miles distant, Manuel's herd was visible, working its way up at a different point — a series of tiny white threads on the mountain. Rosalio had chosen the longer route around, and the rougher one, since none of Dobson's sheep had gone this way and there would be more feed, with possibly a chance for water.

"Harder to travel," he told me, as we commented on our route in contrast to Manuel's, "but more better for the sheep." Personal hardships meant nothing to Rosalio if there would be a benefit for the herd.

Soon after this Rosalio pointed out two lonely saguaros, the last we would see, he said, along the northward trek, and the first of the desert giants to greet them when they would return in the fall. Now the junipers were packing closer and closer until there above they began to cluster into the beginnings of a forest.

Though the morning was less hot than the day before, by 8:30 — three hours after their morning's start — the sheep were dawdling again, requiring considerable effort to be moved. We were working along a rocky hillside around a great scarred branch arm of Borrego Canyon. Listlessness was settling fast upon the herd, as though some sleeping potion dripped on them from the heated air.

Then suddenly a metamorphosis unfolded before us. The drug changed into a stimulant. Rosalio saw it before I did and exclaimed: "See the old ewes run. They go ahead of the yearlings now. Maybe water up there." With a swiftness hard to comprehend after their lethargy, the leaders had already rounded the end of the branch canyon and across it a thin line of them was running back in the direction of the main canyon again. The impulse carried back along that single file of sheep, back to the bunched herd where I watched. Soon a line of animals a quarter of a mile long was running Indian-fashion, leading out from the main body of the herd. It was a sight weird to see, and hard to describe.

The animals at the rear, particularly the yearlings, didn't seem to get the message about the water until it was transmitted to them by the animal next ahead. There was no mass movement of the herd. Each sheep fell in line, as though in a parade, and then ran single file. Some force ahead was pulling out a thin white string from a great hank of woolen twine. Finally the hank was undone; the last sheep at the end of that long thin line was galloping over the rocks and circling the branch canyon.

As the sheep started to run, something had apparently told Rosalio that he would be needed ahead. In a dry country sheep go mildly crazy where water is concerned.

It would have been impossible for him to overtake them. Instead he went down, and then up, the sides of that branch canyon, which was a pageant of wicked boulders and rocks. There he was on the far side, doing the last rocky cliff with ease.

Borrego Canyon had water — small, isolated, scarcely-flowing pools of it — in great cups of smooth hard rock at the bottom, guarded by canyon walls which lowered to twenty feet in height where the sheep entered, but which rose precipitously in height farther up and down. From a high seat on the rim I watched the animals scramble down, then surge out in both directions, for each stranded pool of water was large enough for only a few. Rosalio, somewhere in the distance, was shouting and his calls echoed, sometimes twice, in the canyon. The bleating of the sheep, in its original and in its echoes, was like the booming of a pipe organ, and the red rock ribs of the canyon walls were the pipes.

Later Rosalio joined me. "The sheep run up stream — nearly a mile," he explained. "You hear me yell? I have to turn them back, for that is the wrong direction. I was afraid it would happen."

He had scarcely finished speaking when a different sound came from somewhere below. Rosalio listened intently.

"Hear that baa?" he said. "A sheep is in trouble. I go down and hunt him." (Rosalio always used "he" and "him" when referring to the ewes.)

I heard baas, dozens of them, for the pipe organ swell of bleating was still chorusing the air. But after he distinguished it for me, I did observe that one bleating was not of joy at drinking, nor of anticipation, nor even the call of a mother for its lamb. It was the baaing of distress.

"A sheep is wedged in rocks. Me go down."

Rosalio was ahead of me, over the boulders, until he could peer down into the canyon depths. His call of explanation caught up with me as I tried to follow.

"He no caught. He — what you call it? — stranded."

By now I was at his side and could see it all for myself. At this point the dried-up waterway, leading down sharply, went between walls only three feet apart. There far below was a sheep which, in the craze of seeking water, had kept going along the sharply dropping canyon, leaping to successively lower stages at places where, in time of storm, the water accomplished the descent by low waterfalls. Now the animal had come to the last step in the descending streambed before an eleven foot drop. To the rear the ewe had burned her bridges, so to speak, for, while she had been able to jump down each succeeding fall, she could not jump back up again. The rock, furthermore, was far too slippery and smooth for any climbing. Straight above the sheep for thirty feet was the

87

narrow slit of canyon, then it widened and rose another 200 feet to the top.

And there, ahead and below the trapped animal, was the eleven foot drop, down to a clear pool of water, after which the canyon descended gently, giving a way of escape. I have seen many sheep which would have taken the jump without hesitation but this one would not. It bleated piteously. Rosalio whistled, barked, and threw rocks, but the ewe refused to move.

"I have to go down," he said.

It was a long time, for it required a wide circuit and much climbing, before Rosalio appeared far below, but he found it impossible to reach the sheep's ledge from that angle. He disappeared again and next I saw him, after another extensive lapse of time, working down the slit of canyon from above, the way the animal had come. His shoes and socks were off, for better traction. It was treacherous going. Foot by foot he edged and worked downward, at last lowering himself onto the ledge with the sheep.

In a panic now, just before Rosalio reached it, the animal took the leap and a violent report of noise and a fountain of water filled the air. There was a terrific churning of the pool as the sheep swam the ten feet across it to the rocky ledge. As it came out onto a safe landing and stopped to shake itself, there was a dancing of water particles in the air enough to cause a rainbow, if the sun had been coming that way. In five minutes Rosalio came back up to my position, breathing heavily but smiling. I thought to myself that some herders might have left the animal stranded rather than risk so much in the rescue. Not knowing my thoughts, Rosalio explained:

"One time, when I worked for Mr. Porter, six sheep got caught in this canyon. I sent the dog down but he no could get out too. I had to get them all out."

This was just one more example of what I was beginning to realize — much of the Heber-Reno Trail, for numerous reasons, is too tough for dogs. Often they would help round up strays which lagged behind. But when the herd separated into several branch canyons at once, dogs could not be depended on to make certain all strays were accounted for.

That was Rosalio's responsibility.

Later that day, lingering behind the herd, I found another

pool that was ample in size and only a few inches deep. It spread out in a gentle saucer of a great grayish red rock twenty feet square and the flow was such a tiny trickle that the sun had sponged up all the coldness. Here I took a bath.

At this pool, also, I had companions — such as I had never seen before. One was a green worm, less than an inch in length, with nine legs on either side, and antennae at its head. The second was a tiny creature the size of a navy bean, and elaborately patterned on its back in black and white with the delicacy of a painting. The third was like a shimmering tin submarine, half an inch in length, which resembled the U-boats not only in appearance but in its travel. I am a stranger to creatures of the water but is it possible that, as the desert has vegetation unknown elsewhere, it also may have species of life in its pools which do not exist in other climes?

The imperial thermae of Rome reached a height in bathing grandeur with their library, lecture room, garden, and bathing establishment all combined. But what poverty they showed compared to this bath of mine.

Here was my tub made of rock formed a million years before the Romans lived. Pleated drapes of serrated vermillion cliffs rose about me. Back of those were palisades as grand as on the Hudson River, but more colorful by far. About me was unknown life. Only humans were absent. Few people had ever seen this place. The world felt good this day. Bathing is one of the luxuries.

CHAPTER 15

"Westward I Go Free"

During this trek so far, I have shepherded a dark secret, in the form of some hidden reading matter. Unknown to Pablo, who has to lift their weight in the packs four times a day, and likewise unknown to the burros who must plod along under the extra load — unknown to anyone but myself — I have concealed seven books in my bedroll. Not alone is this probably the first time that anyone other than the herders has traveled this trail in its entirety with the sheep, assuredly it is the first time that *The Private Papers of Henry Ryecroft* has ever accomplished this journey. That volume got included because of its thinness, and since it contained a number of personal recollections which I wanted to study.

All seven of the volumes had size and weight as determining factors in their choice. My *Complete Emerson* couldn't make it. While the thinker of Concord would have lightened many a step for me, he was too heavy — in pounds and ounces — to qualify. William James made the grade, in a compact edition. His *Varieties of Religious Experience* needed reviewing, for this wilderness experience is one variety in itself. I wanted to see if his conclusions led along the same hills and slopes as mine.

The Portable Thoreau HAD to be included. Excerpts from his Journal, some of his essays, and of course his *Walden* had not only been my companions since school days, they were commentaries on this whole wilderness adventure which was unfolding each night and day. His early surroundings had a definite link with this land. To me, the Heber-Reno Sheep Trail is Walden, done in drypoint. Thoreau would have grown gooseflesh in this country.

He needed this trail. It might have meant more to him even than Walden Pond. He was true to Concord because it was what he had, and his whole preachment — his rule of life — was not to bay for the moon. But in Concord scarcely once could he remove himself from sight of a fence. Nearly every acre of land on which he took his walks, although he tried to forget it, was owned or mortgaged, and registered at the county seat, to Baker or Flint of McTavish. To a man of his senses, these facts themselves were fences.

Thoreau needed this trail, and so do those like him today. Although in his time Arizona was still a Mexican domain, and he had never heard the term, still he knew — as a prophet knows — that such a place existed. He needed it. He said so in one of his writings:

"When I go out of the house for a walk, uncertain as yet whither I will bend my steps, and submit myself to my instinct to decide for me, I find, strange and whimsical as it may seem, that I finally and inevitably settle southwest . . . The future lies that way to me, and the earth seems more unexhausted and richer on that side . . . Eastward I go only by force; but westward I go free . . . We go westward as into the future, with a spirit of enterprise and adventure. Every sunset which I witness inspires me with the desire to go to a West as distant and as fair as that into which the sun goes down."

Just as Walt Whitman, in his poems, wrote about western scenes which he had visited only in fertile imagination, so Thoreau — in scattered flashes of his journals and jottings — created pen and ink pictures of this trail, or expressed yearnings which could have been satisfied here better even than at Walden. During our midday stops, I often marked passages from *The Portable Thoreau* which seemed to reflect aspects of this trail experience.

To begin with, he understood the value of wilderness, as we see from statements of his such as these:

"In wildness is the preservation of the world."

"We need the tonic of wildness . . . At the same time that we are earnest to explore and learn all things, we require that all things be mysterious and unexplorable . . . and unfathomed by us because unfathomable . . . We need to witness our own limits transgressed, and some life pasturing freely where we never wander."

"I seek acquaintance with Nature, — to know her moods and manners. Primitive nature is the most interesting to me . . . I wish to know an entire heaven and an entire earth."

"Our horizon is never quite at our elbows."

"I love a broad margin to my life."

"I long for wildness, a nature which I cannot put my foot through . . . where the hours are early morning ones . . . and the day is forever unproved."

Also, Thoreau would have valued this trail experience because of the exposure it gave to these early morning hours. His writings are filled with expressions revealing his love of the dawning day:

"To him whose elastic and vigorous thought keeps pace with the sun, the day is a perpetual morning."

"There are from time to time mornings . . . when especially the world seems to begin anew . . . The world has visibly been recreated in the night. Mornings of creation, I call them. . . . It is the poet's hour. Mornings when men are new-born."

"We must associate more with the early hours."

"Measure your health by your sympathy with morning and spring. If there is no response in you to the awakening of nature . . . know that the morning and spring of your life are past. Thus may you feel your pulse."

"We must learn to keep ourselves awake. Not by mechanical aids but by an infinite expectation of the dawn."

"Only that day dawns to which we are awake. There is more day to dawn. The sun is but a morning star."

The creative mystery of this trail's surroundings would likewise have impressed themselves deeply on Thoreau, as his writings reveal:

"I am affected as if in a peculiar sense I stood in the laboratory of the Artist who made the world and me. He was still at work, with excess of energy strewing his fresh designs about."

"Every day a new picture is painted and framed, held up for half an hour, in such lights as the Great Artist chooses, and then withdrawn, and the curtain falls . . . And now the first star is lit, and I go home."

"I wish to hear the silence of the night, for the silence is something positive and to be heard. I cannot walk with my ears

covered . . . I must hear the whispering of a myriad voices . . .
The silence rings; it is musical and thrills me. A night in which the
silence was audible. I heard the unspeakable."

"The perception of beauty is a moral test."

"What a world we live in!"

CHAPTER 16

Woolly Chess

Our camp for the night was spread on sloping terrain just below the summit of the Sierra Ancha. (Herders usually pronounce Sierra Ancha as "Seriancha." The nearby Mazatzal Mountains, in which Reno Pass is located, are called "Matazals.")

The climb up had been hard for me; upon reaching the campsite, I found a tree for a prop and rested, leaving Pablo with my usual chores of wood gathering. Far below, as supper was cooking, we watched Rosalio leave the sheep and head toward camp. His great strides carried him upward with much less effort than the ascent had cost me and he came into the circle of the fire with a smile and a nod. His first task upon reaching us was to pour a drink for his dogs.

As supper was prepared, and as we ate it, there began to unfold for me the pattern of wizardry with which Rosalio handled the sheep. Throughout much of this trek he was playing a massive game of chess with his herd, and here now was unfolding one of the subtlest moves. It had been 3:00 in the afternoon when Rosalio had directed Pablo where to place this camp for the night. Three hours later Rosalio had left his animals, still grazing and traveling slowly onward, and had joined us for supper.

While I watched him climbing toward us, I also watched his sheep moving along in thin strands, not in the direction of camp at all, but at right angles to it. While we ate, the sheep, far below, were still moving in a direction at variance to the manner in which it seemed to me they should be traveling.

Darkness came and Jupiter, befitting the king of heaven, led the parade of night across the sky. Last night the moon had stepped first from behind the horizon but tonight Jupiter took command and marched at the head of the parade. As the planet and then the moon swung into their stride, they lighted the other marching columns down below — the thin strands of moving wool — and I began to see the strategy of Rosalio's move. The herd was still not heading toward camp but they were progressing toward an impassible barrier, a stern palisade of rock which stood like a Chinese wall before them. It would turn them inevitably in our direction.

Our camp was under a juniper, at a spot in the mountain slope which seemed level. But it was level only in comparison with the sheerness of the drop below us and the ascent above. Actually we were on a stiff grade and all three of us shifted and rolled and tossed most of the night, using our energies in maintaining a sleeping position rather than in sleeping. As I squirmed about in my bedroll, some sand and other debris got into it. When I developed an itch, I wasn't sure whether it was foxtail grass, cactus spines, ant sting, or what.

Somewhat after 10:00 P.M., a soft roar crept through the night — the eerie sound of crunching jaws and of some 6,000 hooves scuffling along the rocky slope. The herd had arrived and, splitting around our dying fire, they moved around and past our camp, creating a nocturnal commotion like night raiders in a battlefield. The animals streamed on either side of my bed. Startled at seeing me, they stopped in the moonlight to stare at me before heading on. Rosalio got up and with his flashlight went ahead of the animals, circled them, and in about half an hour had them bedded down. He brought the herd at last to rest a few hundred feet from where we lay, thus ending a move in this game of woolly chess which he had set up and planned seven hours before.

Forty-six years of experience were behind his strategy; an exact knowledge of the terrain, which would inevitably force the sheep into this one course; a perfect comprehension of the vegetation, just sparse enough to keep the herd moving in search of feed. He had to understand that his animals were hungry and thirsty enough not to stop at darkness as they usually did; he had to take into account the factor of heat which would keep them traveling long after their usual hour of rest. And it all had to be planned so that, when the final stopping place was reached at last, there would be sufficient forage so that the animals could be made to bed down. All that was lacking was a level spot for spending the night.

The bobbing flashlight danced its way back to camp; Rosalio mumbled to Pablo some words that the sheep were down, and once more sought the covers of his precariously sloping bed.

Soon he and Pablo, and the dogs as well, were sleeping soundly once more. And just ahead, the sheep were resting and quiet, a gray patchwork quilt spread on the mountain slope in the moonlight.

But I did not sleep. It was not alone the crazy slope of my bed, but a stirring ecstasy inside me at what I was experiencing. The moon and planet now seemed closer to each other in the sky as they climbed away from the horizon. I seemed to hear Jupiter lean over and whisper to his beaming companion:

"God's heaven, what a herder. He reminds me of that boy David we used to see on the slopes of Palestine. Not since then has there been a shepherd like this one."

CHAPTER 17

"EXTRA!"

I used to wonder how I could go for days without reading a daily paper, but here on the trail I am finding substitutes enough. The red-tailed hawk starts if off of a morning by shrilling "Extra" in its squealing whistle and I have only to turn back the covers — not even go to the doorstep — to start reading the day's events. The weather report comes first, of course; at home I always start by reading this.

That is the way my newspaper here begins. All about me are signs — as adequate as the weather reporter supplies — which tell me if I should pack my sweater, or wear it. The snap of the stars (for this paper of mine is delivered early, even before the news is made), the presence or absence of clouds, the whimsical interpretation of signs ventured each morning by Pablo. All this disposes of the weather and the percentages of accuracy are about as good as the guesses in the papers at home.

It is usual, next, to turn to the comics. Well, nothing is more amusing than Pablo's morning search for the burros, or later, those same burros as they come into camp, trying to eat the labels off the tin cans.

The sports pages of my Newspaper of the Trail are filled with record-setting events; a deer doing a standing high jump which would qualify for the Olympics, a jackrabbit turning in a 100-yard dash that breaks the Olympic record, even the tiny ant hoisting and carrying more weight for its size than the world record weight lifter.

By the time I have perused the weather reports, comics, and

sports, I am ready for the real news of the day — the always changing excitements and episodes of the trail. Of late it has been rare when there was not need for an "EXTRA!" This day called for even more — a special edition made necessary by unusual events. We reached the summit of the Sierra Ancha and came upon a treasure chest of wonders. The Sierra Ancha summit wore a crown, a royal tiara of forest growth — pinyon, oak, yellow pine. But jewel of the crown was manzanita.

Manzanitas. "Little apples." That is what they are termed by the Spaniards, because of the tiny fruit, which was used by the Indians as food and for the making of cider and vinegar and stronger drink. For me, here was a forest of wonderment, a whole mountain plateau woven with tangled beauty, for the trunks and branches of the manzanita shrubs are twisted shafts of stained and polished mahogany, grotesque shapes of reddened, burnished gold.

Reminiscent flashes came to me of my sculptor-artist friend back in California. He sought out rare shapes of the manzanita growth, trimmed and polished them, and sold them as art objects, at prices such as only rare objects of art could bring. My eyes widened; in half an hour I saw half a dozen pieces that might command small fortunes.

"Get thee behind me, Satan." That was no longer the world I was in. Deeply I took in great breaths of the mountaintop air, and began to feel the thrill of just being here, able to witness and absorb all the wonders of the Sierra Ancha summit.

Each year the manzanitas shed their bark and become forest princesses, clothed in a skin of smooth and delicate green. But these mountain maidens sunburn easily, and soon there is a rich chocolate glory of smooth manzanita limbs over all the mountaintop.

It was blossom time as well, and each bush held clusters of fairy bells, urn-shaped, delicate white and pink — lilies of the valley on vacation in the mountains. For two hours we wove through and about this wild riot of color, following natural paths most of the time, for the greater part of the manzanita growth was impenetrable. Nearly in gooseflesh at the display about me, I spoke of its grandeur to Rosalio when next I joined him.

"Yes, pretty," came his comment. "The sheep eat it some,

100

but no much. And it's too thick for them to get through. Awful hard on sheep."

The first thought of Rosalio's life was for his sheep.

At times we came to forests of close-set, tightly packed young pines, nearly all the lower branches dead due to self-trimming, and these places were nearly as hard to penetrate as the manzanita. The ground of much of the forest was bare of grass or growth of any kind and the herd traveled fast. Rosalio advised against my effort to find camp alone and I stayed with him closely. We soon became a two-man school in nature lore, he the teacher and I the pupil.

He told me that in fall one yucca-type plant — in bloom now — would have a long fruit, something like a banana, and "dulce," rather good to eat. He pointed out two wild cherry trees and said that sometimes on their trip back in the fall he and Pablo watched the birds picking off the last of the dried fruit.

With the herd we entered a virginal area, devoid of forage, but with the ground spongy with needles from the forest of pines through which we traveled. No herd had been this way for years, Rosalio told me. The sheep went swiftly through the area, breaking at times into a run.

"You know why they go so fast?" asked Rosalio, and he added the answer: "It is because no other sheep have been here — no tracks."

Pressing for a further explanation, I was amazed at the reply: "All the same like humans. Where other men have been around, men like to stay. If no signs of other men, men don't like it. Same with sheep."

Rosalio was fathoming here the fundamental reason for the growth of cities — the gathering of crowds in public places, the reason for the saloon's appeal, the gregariousness of humans. Few people really like complete wilderness, complete solitude.

Approaching Gun Creek another "special edition" event unfolded. The sheep stampeded into a rocky branch canyon in hope of water. Suddenly, under the pounding weight of those hooves, a section of the cliff gave way. Some thirty animals slid and fell to the canyon's bottom, with rock and dirt tumbling and showering around them. The drop was perhaps twenty-five feet and, by some miracle, no casualties resulted. The rest of the herd were able to stop and thus avoid the fall.

Deliberately, yet with resolution, the animals which had not been caught in the slide, and which were still above it, began a maneuver which confirms my belief; sheep have been makers and molders of history.

One of the older ewes began slowly and bravely picking her way down the sheer slope of the debris of the slide, going in sharp zigzags to decrease the grade. It was a hard assignment — and a dangerous one — picking her way around and between rocks as she pioneered a safe path to the bottom. The process took probably ten minutes.

With one safely down, another followed. Then another. In half an hour the entire herd had made it to the bottom. Sheep must be placed prominently among the trailblazers of history.

I have heard that some of Boston's streets follow the meanderings of early cowpaths. It would be interesting to know how many furrows in the world of humans follow the trails of sheep.

We made camp for the night on a gentle slope among junipers. As we were drying the dishes following a very late supper, the herder from the band of sheep ahead of ours dropped in for a visit. He had lost his campero and so, bedding down his animals, he had walked several miles through the darkness in search of our fire. This man ate with us, then headed back through the wilderness again, to be with his sheep during the night. It was cold now, on top of the Sierra Ancha, and he would be without camp or blankets.

Later we found out that this herder and his campero had been at odds with each other most of the trip. Trouble between them, we discovered, was brewing.

Thinking of the manner in which the varied events of the trail resembled the format of a newspaper I drifted into sleep. "That's 30 for tonight," I thought.

CHAPTER 18

Gathering Storms

The trouble in the camp ahead is causing us to slow down, for our sheep must not catch up with theirs. So we decided to make our night camp serve as well for the noon stop.

Rosalio came in at 10:00 A.M. — having quartered his sheep in fair pasture on a juniper-clothed mountain slope some distance out — and ate lunch of dry bread and mutton stew. (We had mutton only when Rosalio discovered a sheep lost from some other herd or when one of our own animals became too sick to continue.) Then he journeyed back across the canyon and rocks to be with his sheep for the day, taking pencil and paper with him to sharpen some dull moments writing to his sister in Old Albuquerque. Of seven in his family, just Rosalio and his elder sister remain. Four times a year he sends her a letter, and goes to see her once a year.

This day started cloudy and it has steadily worsened. Now it has developed into a manner of weather such as we have not had before on the trail. There are cold, formless clouds pushing down on the Sierra Ancha, gray gusts of wind (gray in spirit if not in hue) and a melancholy rain is starting. At 3:30 P.M. — just now — Rosalio returned to leave his finished letter in camp and to take his raincoat out onto the hills as he watches the sheep. It is an old coat, worn through at the shoulders, but a shiny new slicker will be brought up to him for summer in the White Mountains, where rain is frequent.

This is the kind of day that sets one longing for shelter. I had chosen my bed — employing it also as a writing table and reading stand for the day — out by a fine slab of red sandstone in the clear some distance from camp, with unobstructed sky above and forty miles of mountains sweeping below me. Ordinarily I do not wish the branches of any tree barring me from the stars at night. But now the thick-grown needle clusters of the pinyon close by camp have a look of warmth and shelter about them and if the rain intensifies I shall move my home up under their protection.

Toward evening the rain set in — not heavy at first, but one shower after another. Pablo groped through the pack boxes for his almanac, unused now for several days. It is printed in Spanish and quotes the weather for the Southwest and for Old Mexico. For this particular day it said for Arizona "dry, with dusty winds."

"Well, maybe she right," observed Pablo. "Maybe this rain blow over pretty quick." By now it was 5:00 P.M.

But each gust of rain became worse. Pablo had his black slicker on by this time and I had not only moved my bedroll up under the pinyon but as the rain came I used a tarp over my shoulders. The camp, even though the showers were light, even though the forecast said "dry," obviously was getting wet. Again Pablo consulted the almanac and read me his new findings.

"I think this rain comes from the moon changing. I see it changes quarter in three days. But hell, three days one way or the other don't make much difference to the moon. I think I put up the tent."

A thin canvas herder's tent, just large enough for two bedrolls, and with iron stakes and pole, were part of the equipment carried by our seven burros. It was really little more than a hanging tarpaulin.

Pablo and I set it up, though it was nearly impossible to drive the stakes. An inch under the floor of our camp — two inches beneath the soft bedded needles of the pine — was solid rock. The earth three feet from where we pounded showed movement with each blow, indicating great slabs of sandstone such as came to the surface in the ledges not far away. Tying one corner of the tent to a bush, and another to our heavy water keg, at last we got it up. With his lasso Pablo roped the projecting tent pole up above, and guyed it to a tree.

The showers steadied into continuous rain. At nightfall Rosalio came in, having driven the sheep into a gorge just below camp. But the animals would not bed down, so while Rosalio ate, Pablo made his way into the canyon and drove the herd close about our tent.

Darkness came fast, with just a whisper of sunset color in the west. Pablo and I had hauled in three great logs — two oaks and a juniper — which would have cost a day's wages in Phoenix, and these we fed to the fire one upon the other. Even as it rained, the flames roared up and warmed us. I spread three heavy tarps over my blankets, the others laid their bedrolls in the tent, and we said "buenos noches" early. While adjusting myself to my blankets a heavy shower came and I pulled the tarp over my head. The tapping fingers of the rain, just above my ears, tattooed a nervous melody on the canvas.

The urge to see what was transpiring made it impossible to stay prisoner beneath the tarp for long. I looked out. The expensive logs were still burning briskly, though the rain on the bed of coals sent puffed swirls of smoke up into the air. It was tinted smoke — great luminous billows of it — lighted by the flames beneath, and it carried a bit of cheering glow up into the blackness.

Now the rain bashed hard and pounded into my face. For a minute the sensation was good — a cold spray pounding my cheeks in a hundred places. I removed my felt hat, for it was part of my bedclothes this night, and let the drops sting my head. I loosened my sweater and they needled my chest. The juniper log, prodded into revolt by a blast of heavy rain, blew a smoke ring of cedar incense out into the circle of the camp. Its strong tang was as tingling as the rain. I pulled the tarp back over my head, capturing a billow of the tang within my bed. It drugged me to sleep.

That was the last time I could look out during the night. When, somewhat later, I was wakened by a hard beating against my canvas prison, the storm had ceased its playing and a cloudburst had swirled into our mountain home. The tapping above my ears became an urgent violent stamping. The white flare from flashes of lightning easily penetrated my cover and the roar of thunder, following close after, to tell me that the center of the storm was near, seemed even magnified by the tarp, as though the canvas were a sounding box. Donner and Blitzen were on the march. With the lightning coming more often I wondered if I had chosen wisely to seek the tree for shelter. Rosalio had told me that thirteen sheep, of another herd, had once been killed by a single bolt of lightning.

The tarp at my feet blew away and the bed down there began sponging up water; some rain ran in by my head. But I would rather be a bit wet than to attempt fixing things, for fear of making them worse.

At some time or other I dozed off and must have slept soundly for the remaining night. For next I knew, Pablo was cutting wood. Rain was still coming down but he was up, the fire was freshened and roaring, and hot mush and coffee were nearly ready. I rolled out and slopped into my shoes.

CHAPTER 19

Heroes in Mud

Today I learned the stuff which makes a herder.

By 5:30 in the morning when breakfast was over, the rain had established itself into a dismal storm. And the air was swiftly chilling, turning cooler. There would be a cutting downpour for half an hour after which it would cease a few moments, then set in again.

The sheep this day, far from requiring less care, would be doubly hard to manage. Already we had spent a full day here and the surrounding forage was exhausted. Until we could move our quarters Rosalio would have to circle his herd far out and back, across the canyons, seeking feed. Not even the dismal, drenched shelter of the fireside would be his. Breakfast over, donning his worn-out slicker, whistling to the dogs and then to the sheep, staff in hand, he set out. The soaking hills and dripping trees, the fury of the driving rain, soon swallowed the herd and herder; they were in a woeful world.

Pablo's lot was scarcely better. It was urgent to move camp if the rain slackened and even though this chance seemed remote, the burros must be ready. On went his slicker too, and he went out into the rain to find them.

I remained in camp and piled wood upon the fire. The weather, wicked now in its moods, held an exciting attraction. While the rain was pelting down, a grip of coldness clutched the mountains and the raindrops turned to hail. The stones bombarded the tent; they lit in the lids of the Dutch ovens, hot over the fire, and burst into pockets of steam; they danced sharply on the brown slick slabs of sandstone rock near the camp; it was a dance of bitterness,

107

fast-rhythmed and striking a chord of pain.

Gradually the storm subsided but even though the rain slackened, the clouds began to push down upon our heads. Some of them were raking the mountain spur to the west and a thin finger of cloud, looking like fog, drifted into the valley adjoining us. Over above Potato Butte in Pleasant Valley the fog was playing blindman's buff, more blind than buff, for soon its cold white clamminess stole into our camp and for five minutes the world was gone. We were submerged, cut off, grayed out.

But that mood also passed and the rain began again, although now it was almost continuously hail which came at us like icicles. Our mountain slope became a rock-ribbed refrigerator. The tent became the coldest place of all, for its flaps were a foot above the ground and gusts, which nearly tore the whole tent down, started up and swirled the shelter full of icy air.

Pablo returned with the burros and his horse, and when the rain ceased a moment the horse began to steam, its brown flanks — streaked with wet — smoking as though on fire. The burros cut the sorriest scene of all. A burro is never happy, at least in appearance, and now each animal was a document of woe. Like silent statues, they stood there, wet and dripping, their ears hung back and down like the tails of whipped pups, and the water streaming off each eavetroughed ear. A burro's ears are his only glory — the only flag of pride which he flies — and now these had gone to half-mast. When the rain got too unendurable the men swore, the dogs shook themselves, the sheep bleated. But the burros just stood, and let water run from their ears.

Rosalio returned at 10:30 in the morning and we tried to eat, although the rain made it a quick and soggy affair. Then Rosalio tried to dry out, for the slicker had leaked badly at the shoulders and his overalls were soaked from the knees down. He had been in mud and his shoes were both wet and clay-packed. He knelt by the fire to scrape off the mud with his knife.

After that, Rosalio and Pablo and the two dogs went into the tent to rest. The dog, Georges, was most depleted of all. He was chilled and couldn't get warm, and shook violently for an hour. Pablo covered him with a pack blanket, then curled his own body close to the dog and they both slept.

A fury of wind and rain caught us in an instantaneous grip.

The dams and ditches which we had constructed about the tent began to break and Rosalio and I repaired them. Hail followed the rain and came with such volume that it bounced under the flaps into the tent and melted in large pools of water. This we could do nothing about. Pablo and the dog slept through it all; they were exhausted.

Rosalio and I, in a lull, sought the fire and piled more logs upon the diminishing flames. We had burned seven so far during the day; our one piece of luck was being close to good fuel.

Over along the mountain spur to the west we could see a storm — the worst of the day — lashing its way along like a Paul Bunyan gone mad in the forest. In this country you can see your enemies coming for many miles. To the east was another storm, less only by degree — Paul Bunyan, Jr. — traveling in the same direction. They both skirted our camp. We were happy in the thought they would miss us. But the tapering ends slowly closed in. Ahead of those ends — a circular fury between the two storms — came wind. It tossed and swayed the tent; from my bed it lifted the tarps, stones and all, with which I had weighed them down, carrying them to the other side of the camp; it tossed fire embers up like crimson popcorn. Then the two ends of the storm met and fought it out with our camp as the field of honor. It was a family feud; a battle of incest. I do not know which one won, but we lost.

Rosalio had driven the sheep in close to camp but now, at 1:00 P.M., they became restless for feed. He picked up his staff, put his wet slicker over his clothes which hadn't yet dried, and started out again. For six hours he would drive the herd in great circles through the mud and hail and hell. I watched him go; under similar circumstances I question if I could have done the same. As the figure of the sixty year old man headed out toward the sheep, another figure slipped from under the tent flap and dashed along behind. Rosalio had not whistled. But Georges — still shivering, still wet — would be out there too.

It rained and hailed, with short letups, all day long. I pulled in five more logs from the forest. With slabs of slate rock pried from the ledges, I constructed sidewalks about the camp, for the pools of water and mud now made walking a chore. When the rain slackened enough so that the prospect of drying out was a little greater than that of getting wetter, I stood by the fire. The denim of Levis in some respects resembles sheet iron; my pants would get too hot to touch and then, in stepping away, I would bring them in contact with my legs. There would be a stinging burn.

During the worst storms I lay in the tent, watching the rain-drops race along the bottom flaps, then drip to form pools where the beds would be. Across on the far hills I could hear the bells of the sheep. Rosalio was out in it all.

Pablo had cooking to do. Worst of all we were nearly out of bread. Half an hour before, he and I had eaten the last soggy slab, throwing it first down into the ashes to drive out some of the damp-ness. We only succeeded in burning it; it was still soggy.

With a hope that the weather would change, Pablo delayed his baking as long as possible. But the rain continued, so he got out his pans and his flour.

I don't know that I have seen a sight like that, ever; this Mexi-can-American campero, swathed in his black slicker, his hat drip-ping, mud clinging to his feet as he sloshed about the camp, mixing his dough in the rain and the hail. His partner was over on that for-ested mud-sloughed hill, drenched now through and through, mis-erably cold, yet tending the sheep. That was Rosalio's job. Pablo's job was to have bread ready when his partner came to supper. He wasn't going to shirk it.

It was bread being mixed in that pan and being baked on that fire. But as I saw the brown loaves coming out, the very sight of the bread told me it was more than baking dough; I realized here was real proof that people do not live by bread alone. They can rise to great heights even in a sheep camp.

Now it is 5:00 P.M. I am in the tent, watching the rivulets of water break the dikes and run farther and farther in. I have just remembered that four of Dobson's herds as well as Mulatto's and Manuel's are out in this same storm. The men of those camps are suffering as much as we are. Manuel and Eduardo, his campero, are probably suffering even more. They are fighting not only the weather, but themselves. A storm within can always beat the fury of a storm without. (We learned later that it snowed in their camp.)

Tonight will be bad, I fear. Last night's rain was light by com-parison and the weather was still mild; now it is clawing, cold, and windy. Cold and wind are foul handmaidens of rain. Last night's rain had some poetry in it, at least a stanza or so; tonight's, I'm afraid, will be all prose.

The trials of this day have been much on my mind. To me it has been hard, but I will be through with it soon. Rains like this, however, occur all summer in the range of the White Mountains where the sheep are headed. The herders will be wet through them all.

There are other workers who must labor in rain, who are soaked through and through. Police on their beats, and lumberjacks, those who deliver the mails, and farmers sometimes, and the miners "working wet" in dripping stopes underground. I did that for three months once. But all such persons labor for eight hours, or twelve at most, then have quits. There is large comfort in the thoughts of the fire and home which await them. Rosalio, over there in the drenching hell on that hillside will, at nighttime — just an hour or so from now — have to exchange that particular hell for one nearly like it in camp. More cold, more wind, and a sodden bed. The only fire is out in the rain. Any warmth he can enjoy will be at the cost of a soaking. And through the night the sheep must be constantly on his mind. It is nearly as bad for Pablo.

I am wondering this day that sheep tenders can be hired at all. Not eight hours, not twelve, but twenty-four, and each one sixty minutes of torment in times like these. Not five days, nor six, but seven including holidays. It is marvelous what men will face and bear up under. And I have heard people call sheepherders lazy.

Now it is nearly night. At supper time the rain slackened a bit and for half an hour there were signs of clearing. The coyotes yelped and Pablo took it gleefully as a prophecy.

"See, they over on the hill yelping. They are happy and make big dance. It happens every time. When they yelp it means a change of weather."

I was becoming quite taken with Pablo's coyote theory until, just a moment ago after it had started raining some more, the coyotes bayed forth anew and Pablo said: "It cleared a little but now it changing again, back to rain. The coyotes yelping means another change." He was hedging like a true prophet.

The Spanish almanac said "dry, with dusty winds" for today. It was only two-thirds wrong. It was windy, I can't deny.

111

CHAPTER 20

Shots in the Night

We camped under a large juniper near a rocky cairn, between Charcoal Mountain and Spring Creek, on a high plateau of the Sierra Ancha. Rosalio roused us at 4:30. The sun topped the mountains, and the sheep voluntarily started at 5:45.

Rosalio has just left with the sheep and Pablo has gone to look for his burros. He didn't know whether they went in the direction of the ridge ahead or down into the canyon at our right, so had to track them. He apparently found them on the ridge, and now I hear their bells approaching. I am here guarding our supplies, for that is one aid I can render. Otherwise one burro might come in ahead and start eating up everything in sight. When Pablo is alone he has to pack and cover things well.

Juniper makes good firewood and the smoke is incense — although Pablo claims that pinyon is better. Rosalio spent a full ten minutes this morning covering the fire with rocks and dirt — and it was hard shoveling dirt in this rocky land. He not only wanted to put out the fire but douse all smoke. I wondered at the caution — far beyond ordinary needs. He explained: "Don't want Pleasant Valley ranger to see smoke. Then he come and see my sheep tracks outside of line."

He was referring to the fact that last night he had driven his sheep outside the established boundaries of this Heber-Reno Stock Trail. I asked what the penalty would be. "Jail, maybe," Rosalio responded. Although I doubt if the penalty would have reached any such proportions, he had been willing to take risks so his sheep could feed properly while stalled by other herds ahead.

Now the burros are here; the thump of their feet as they jump in their hobbles is mixing with the sound of the bells.

Starting out, Rosalio and I soon met our caporal, Crescencio, whom we had not seen recently, and found that the trouble in Manuel's camp is worsening. He and his campero — Eduardo — have been fighting, and Eduardo is going to quit at Pleasant Valley.

At noontime Rosalio went to eat in Manuel's camp and Pablo and I ate alone. After our simple meal, Pablo cleaned up, put some beans to cook, then saddled his horse to ride over and see the men in the other camp. He has left me here alone to watch the fire, to add water to the beans when they need it, and to keep the burros away.

I sit here, the whole forested meadow to myself, the vast mountain range to myself, the whole world to myself for all I know or anyone else knows and for all anyone can do about it. Few times do I find myself in a situation just like this; I am lord not only of the sheep camp but of all the Sierra Ancha.

Suddenly I realize I have stumbled on a reason why many men become herders of sheep, or camperos. Every day he is alone in camp Pablo would have this same feeling that I have now. It could never come to a worker at a lathe in a factory, to a clerk in a dry goods store, to an employee anywhere with fellow workers about and with the boss up front. It could not even come to a sailor working under his mate at sea. But here — even though he is employed — Pablo is not only his own boss but he is lord of all he surveys and king of the hills. Freedom! He is not a little man. He amounts to something.

The beans were bubbling merrily. They were our entree for supper when Pablo and Rosalio returned.

In the night we heard two shots — obviously from Crescencio's rifle, for it was the only one around — off in the direction of Manuel's camp. During the evening we had been discussing Manuel, and both Pablo and Rosalio agreed that he had been on the trail too long — the stress was too great — which was the reason for his surliness and the fights with Eduardo. The desire for Eduardo to leave was the fear that Manuel might harm him. And now these shots — two of them in the night; I half imagined, half dreamed, that Manuel had gotten Crescencio's rifle and used it on both the

114

campero and caporal.

I didn't sleep too well.

After Rosalio and I had started out in the morning with the sheep, Crescencio came along and told us the cause of the shots. In the night, the sheep had become frightened — and the burros and horses as well — either at a bear or a coyote, and started stampeding. Crescencio had fired into the air, either to frighten the wild beast, whatever it may have been, or to stop the sheep. I am not sure which. I doubt if Crescencio is sure either. Rosalio, with no weapon at all, controls his herd much better.

Crescencio as caporal, or foreman, receives higher wages than a herder or campero. Pablo is envious neither of the caporal nor the herder. A caporal, he says, may not have to work quite as hard as a campero does, although there is much to do — aiding in counting the sheep, riding ahead in search of feed and water, helping in other ways. "But he has to deal with the men in all three camps and I don't like that," Pablo explained. Here was one worker who realized that the person above him had a job of responsibility and, because of that fact, was entitled to more pay.

Pablo likewise knew that Gunnar Thude, the owner, and his associates would make a handsome profit from the lambs this year, but he also knew, not only the expenses involved, but the risks. "You got to have a gold mine to go in sheep business," he elaborated, "and if you not careful, you go broke sure as hell."

For several days the talk has been about Spring Creek, as though it were some ogre vaguely looming ahead. Although there has been talk about it, we arrived at Spring Creek Canyon with no immediate warning at all, so far as I was concerned.

In terms of beauty it is a Garden of the Gods; in terms that the herders use it is Hell's Half Acre, though they call it by other words. Actually it is hundreds of acres of rock-strewn, canyon-scarred tumbled madness.

I myself am not a swearing man; I would not blaspheme the burros as Pablo does — though I certainly don't object to his doing it, and think it actually lubricates the packing. I would not swear at the sheep, as Rosalio does on rare occasions and with just cause, when their true sheepishness comes to the fore. Neither would I damn the weather, as all herders do, though I have been chilled by it, soaked by it, baked by it, stung by it, so much already on this trail, that I might be excused for some original words in its regard. Nothing ordinary can raise me to profanity — or lower me to it, depending on one's slant and frame of mind — though I who seldom swear like to think that there is a language of abuse somewhere beyond the ordinary tongue, to be drawn on for special occasions. Those who curse at the scratching of a fingernail, or profane the slightest annoyance, are not getting the true essence out of pungent words; they are cheapening what might be glowing terms. If one saved the vocabulary of profanity for those memorable times when it really should be used, then I think it would become an honored language.

In all the days and in all the trials of this journey so far, I have used but one full-flushed oath. As I crossed Spring Creek and turned back to look at the whole area in its viciousness, I gave vent to almost reverent profanity. "It certainly is a !-*-#-@-*-@-*-?-!-!," I exclaimed. At the time, there was no other way to describe it.

Spring Creek is simply a series of canyons whose rock-ribbed walls rise grandly in towers standing against the sky. At a casual glance it is not impressive at all. But to one who goes through it with fifteen hundred sheep he finds that it has just that roughness, that sheerness of trail, that confusion which causes all but the best to get lost, which makes it worth an oath. In terms of hazards, it is the climax of the Sierra Ancha.

Also, it is the gateway to Pleasant Valley.

116

CHAPTER 21

Herders' R and R

Pleasant Valley is logically named — and strategically situated. To Pablo and Rosalio and the herd it was a place of rest and refuge between two of the cruelest sections of the Heber-Reno.

Back of us were all the tortures of the trail leading out of Tonto Basin, the weariness of threading through the maze of manzanita on the Sierra Ancha plateau, the canyons and rocks and treacherous ledges, the fury of the rains which held us prisoners in misery when we had started down from the top. In the Sierra Ancha, within a week, the weather had gone from rain to hail, the thermometer had tilted from well above 110 degrees to freezing. It had sucked the mercury up to the heights, then plunged it down to the depths. Cruelest blow of all had been Spring Creek Canyon, that hellish nightmare of rock-strewn, gorge-scarred tumbled confusion, just before we came out into Pleasant Valley. All of that was now behind us.

But ahead lay the three devils of the Heber-Reno — Naegelin Rim, Ramer Canyon, and the Mogollon. Here in Pleasant Valley was the chance to repair from what was behind, and prepare for what lay ahead.

After drifting idly for a couple of days across the grass-blanketed valley floor, Rosalio chose a camp spot surrounded with lush forage and shaded with a few scrub oaks, close to the base of the Naegelin. The night before, we had done little else but relax. On this afternoon and evening we did nothing else but prepare for the assaults ahead.

Carefully Pablo inspected the shoes of his horse, and

replaced one. From his pack, Rosalio took out a newer pair of shoes for himself than the tattered ones he had on. Adjusting his glasses, which he almost never used, and whetting his pocketknife (every good herder carries a small whetstone) he cut out the upper leather from an old shoe and pared it, by a great circular cut, into a rawhide shoelace, three feet long.

At a deserted trapper's shack below Spring Creek, Pablo had found an old washtub. "Give me your shirts, fellas," he commanded, and the three of us stripped to the waist. After Pablo's laundry had been boiled in the tub, it soon adorned the scrub oaks like flags at a carnival. Then Pablo spent nearly an hour scouring every pot and pan he had. Twice he arranged and rearranged his spices and canned goods, as fussy as a newlywed bride. From somewhere in the duffle, Rosalio produced a saw and hammer and skillfully made repairs on a kyack box, which had broken.

When no more work could be found, Rosalio searched in the packs for his battered mouth organ, which he had used only three or four times on the trip. He is an expert on the instrument. Mexican music and a couple of lively Yankee airs floated into the breeze, mixing with the juniper incense. Using a recently laundered dishcloth as a veil, Pablo did a spritely dance.

As Pablo began packing his pots and pans, I thought of all the meals he'd prepared, and marveled at his skills in these kitchens of the open air. There is a science to cooking over an open fire and Pablo knows it well. It isn't the way of the Indians, who need little wood. When he has it, Pablo uses much, and of the best kind to produce embers.

He has three white covered baking pails or kettles and two aluminum ones which go through the whole trek without once being streaked with black.

He usually digs a hole for his fire and fills it with wood until the hole and its fringes are filled with glowing embers and ash. Then, scooping some of these in a shovel, he makes a patty of hot ashes on the fringe and places his pot or kettle on this. It does not blacken any more than on an electric grill. Not heat, but smoke, smudges kettles.

Much cooking is done in the Dutch oven, an iron kettle with a lipped lid into which live embers can be placed, so that food cooks on both top and bottom. Coffee is always made in the Dutch oven, then poured into the coffeepot. The pot never goes on the fire itself, but is kept on live embers.

Late in the evening Rosalio again filled the air with south-of-the-border tunes from his mouth organ. I fell asleep to the music.

Our morning start next day was earlier than usual for we could not make noon camp until we had scaled the Naegelin Rim. The grassy slopes and scrub oaks of Pleasant Valley soon turned to thick underbrush. I started up the slope ahead of the sheep — scrambling over sharp, loose stones — and finally found a large opening in the Rim up which the sheep would come. From a pinnacle at one side of the natural gate I could look down at all that was happening below.

At one point, the entire herd stopped. Behind, Rosalio was urging them on. His piercing whistle bit into the air. He waved his hands and his staff. It was a sheer ascent and the herd's forward movement was almost imperceptible.

Then the action suddenly became swift. Wave upon wave of sheep clambered up the rocky slope. Opposite me came the greatest test, in which the animals had to leap up from one ledge to another. The driving was easier now. The herd, once started, seemed to accept the challenge and the goats and leader-ewes found half a dozen different avenues up which they led seemingly never-ending files of sheep. Sometimes a sheep would not make the upward leap and would slip back, then either try again or circle and make the attempt at some other place. Leaping, clambering, sometimes hesitating slightly but never for long, each animal scaled the Rim.

The noise resembled the crescendo of the 1812 Overture — thunderous whistling and shouting of Rosalio, barking of dogs, bleating of the sheep. Just beyond the herd was a sheer face of rock over which even the sheep could not climb. Or could they? Some animal was surely over there. As I looked, a deer — frightened by the herd — took a graceful leap up over the barrier. The sheep leaped too, though far less gracefully. But they made it to the top.

Rosalio followed the stragglers and carried a bleating lamb over the last stiff climb. Almost directly behind came Pablo who, with his pack animals, had followed a narrow ranger trail to the top. As quickly as it had started the ascent was finished. Soon sheep, burros, horses and herders — all of us — were swallowed up in the dense brush and undergrowth of the trail along the top. In a small clearing, we made our camp.

CHAPTER 22

Horseman of Mystery

Ramer Canyon is an enormous gouge in the earth, strewn with twisted pine forests, ravines, and ridges. But hazards to the sheep came far less from the ravines and ridges than from the forests and tangles of undergrowth, which swallowed up almost the entire herd from sight. Sometimes we could scarcely see a dozen sheep at once, out of the fifteen hundred.

After we got started down, the sheep moved fast. I knew from the sounds that Pablo was working one side of the herd and Rosalio the other insofar as anyone knew where the sides of the herd were. Certainly I didn't. Usually a campero does not help with the herding. Now it was a necessity.

But in his endeavors to help with the sheep, Pablo had let his pack animals drift, and Chino — the burro carrying our two water kegs — was missing. Rosalio and I held the herd in a bunch while Pablo turned back on foot. Immediately he was lost from our sight in that snarl of underbrush.

He was not gone long. Chino, still bearing the precious water, had been found standing in serenity back in the forest. "That damn burro didn't even know he was lost," Pablo mumbled. "He was just standing there."

Through the forest we progressed, calls and shouts ringing out as Pablo and Rosalio endeavored to keep contact with one another.

They seldom shouted in the course of their everyday work. But they were shouting now — to each other — for only in this way could they hope to keep track of things.

121

I called occasionally too, and the forest was alive with echoes. The echoing, in fact, began to seem unnatural. Something was strange. Midst our own shouting, a fourth voice was making itself heard. It was coming from down below, and slightly ahead. It was some herder from another band. He was yelling to us — yelling a warning. Our herd was heading into a large canyon. This other herder's sheep were coming up this same canyon, to a junction with ours. Surrounded by the impenetrable underbrush, it was as though we were traveling with blinders. Our two herds would mix, which meant they could not be unscrambled again until after topping out on the Mogollon. An awful possibility.

Rosalio sprang into action. Pablo and I did too. Down there ahead of us we knew that the other herder was acting also. We three flanked our sheep and headed them back. Then we scouted ahead and discovered Juan Estrella, with Gibson's herd. He had turned his sheep back likewise. With the animals under control there was time to exchange experiences. Rosalio and Juan agreed on separate routes which they could follow, and we started on our ways. There would not be enough feed for two bands to follow along the same trail.

Toward evening the lead sheep broke into frantic, almost pitiful baaing. Almost at once every animal in the herd took it up. The soft volume swelled and filled all the space beneath the great trees. Pablo rode up on his horse. "It's the dried-up spring by the old corral," he explained.

It was almost dark, so Rosalio instructed Pablo to continue on to the corral and pitch camp. Pablo told me later that the corral had once served cattle rustlers as a place to contain their stolen herds. Rosalio and I followed the sheep until I saw the light of Pablo's fire ahead. I left Rosalio to the task of bunching the herd for the night and started into camp.

A thickly-forested dark canyon led down into the flat on the side opposite camp, and as I approached the corral I saw a rider on a pinto horse come down this canyon and dismount. For a time he stood secluded by a group of great jack pines, peering out at Pablo. I waited silently, expecting the man to head into our camp. But, instead, he mounted his horse and galloped back up the draw.

Going on in, I recounted the episode to Pablo. It was completely dark now and soon Rosalio came in from the sheep. I could catch such words as "pinto" and "pronto," as Pablo described to his companion what had happened. This area between the Naegelin and Mogollon Rims is one of the most remote in Arizona. It was, Pablo said, a hideout for criminals from the cities. No man, we felt — either acquaintance or stranger — would approach so closely to our camp and then turn back without at least a word of greeting. No one, that is, unless he had suspicious intentions.

While we were discussing the event, a small band of sheep — a quick count showed there were fourteen — came running out of the darkness from the canyon into which the man had disappeared. They bore the Dobson brand. Failing in an attempt to corral them, we drove them up to our own herd.

This added strength to the suspicions which had been forming in our minds. The mysterious stranger, we felt, must be hiding down here to cut off small bands of sheep. This might be part of a large black market operation. Perhaps he had cut the small Dobson band and, discovering us, had been forced to let them get away. Rosalio had left our own herd a little distance up the mountain slope but he and Pablo, with flashlights, went out to drive the sheep right down to our camp. During the evening the dogs seemed to growl more than usual. None of us seemed anxious to sleep. It was midnight before we finally turned in, and I chose a spot about fifty feet from camp, out away from the circle of firelight where I could see anything that might happen. The sheep were restless throughout the night; three times they stampeded briefly but soon came to rest once more. But this often happens as a result of coyotes or other scares. When morning came the herd was intact.

The Mogollon Rim is one of Arizona's dominant geographical and geological features — a natural barrier extending for 200 miles across the east-central section of the state, and rising in places a thousand feet above the land below. But to the sheep, after the struggle up the Naegelin Rim, and through the hazards of Ramer Canyon, it was just one more grueling day's work, which they took in stride.

The emphasis would have to be on "stride." Rosalio's long stride carried him up a branch mountain slope where a dozen yearlings had taken a wrong direction. The old ewes did more than stride — they jumped, actually having to leap up to surmount a barrier, or reach a ledge.

It was each animal for itself — individual effort all the way — a strange, unbelievable, heroic sight. While the air was still filled with plaintive cries of the sheep and barking of the dogs, I suddenly realized that we had reached the top.

After topping out over the Mogollon we had time for only a few deep breaths of this purer more rarefied air — the altitude is 7,400 feet — before the Sitgreaves National Forest waved us welcome. On the dirt road a few miles to the east is an official Forest Service sign, but in the still pathways of the forest where we traveled, only the many-fingered pines gestured their pleasure at our approach. No boundary, except on maps, exists between Tonto National Forest which we had entered just before reaching the sheep bridge, near the start of the trek, and this Sitgreaves Forest through which we would now be traveling much of the time until we reached the Fort Apache Indian Reservation in the White Mountains.

Three miles farther on was the Mule Springs Corral, near the little-used dirt road leading from Pleasant Valley to Heber. This forested area surrounding Heber is the summer range for many of the sheep which are driven upward from the Salt River Valley; the corral has been built near the crest of the Rim, so that possible strays may be taken out and returned to their proper owners.

Rosalio and Pablo, driving their herd into the corral, did not find it difficult to remove the strays. Brands painted on the wool on the animals' backs made identification easy. We had picked up fourteen strays from Dobson's herd and a few sheep with other brands.

Three other herds came in on the day of our arrival, and herders, camperos, owners, and hangers-on had a sociable afternoon. The meeting at the sheep bridge, near the start of the trek nearly five weeks before, had been a time of pleasantries; this coming together at the corral above Mogollon Rim was the occasion for tales of adventure and woe.

We learned that one herder, climbing hand-over-hand up a rock-slabbed cliff, had been bitten on his hand by a rattler. He rode horseback into Pleasant Valley, where the ranger rushed him by car over a wicked road ninety miles into Holbrook and medical aid. The hand swelled, turned partially black, but the herder had recovered fully.

We heard the story of one of Gibson's herders who, down in the vicious wilds of Ramer Canyon, had lost his campero who had recently joined the trek and was unfamiliar with the country. Five days later, hungry but unharmed, the man was found by a deputy sheriff.

Among the herds that arrived at the corral during the afternoon, there were a dozen or more sick sheep, and two sick horses. One of the Dobson herds had had a cut or loss of a hundred sheep, and one herder "sheepishly" admitted that he was nearly 200 short. At this meeting at the corral, nothing was related to equal the tragedy of four years ago, when one of the herd owners, taking the place on the trail of a herder who had become ill, had his eye put out by a cactus spine. I said that this gathering at the corral was a sociable occasion, but in some respects it was more like the waiting room of a hospital. Men and animals had been worn down by the tests and trials of the desert and the Sierra Ancha. Now the Naegelin Rim, Ramer Canyon, and the Mogollon had delivered the final punches.

Rosalio and Boots. Much of the rugged trail was hard on dogs.

Upper—Peals of thunder cause the herd to huddle in fright. Lower—As they start their descent into Spring Creek Canyon the blackening storm makes progress dangerous.

Upper—From restful Pleasant Valley Rosalio eyes Naegelin Rim which the herd must surmount. Lower—The sheep fan out in forested Ramer Canyon, another hazard between the Naegelin and Mogollon Rims.

Upper—Rosalio tends an injured sheep after the strenuous ascent of the Mogollon.
Lower—From top of Mogollon Rim, author looks out over route just crossed.

Upper—Cool forest pools provide ample water the herd needs after journeying through parched areas.
Lower—Rosalio always sees that his dog is never thirsty.

Upper—Rosalio often helps Pablo with meal preparations.
Lower—In the cool forests, Pablo has a good supply of wood for his campfires.

The arduous trek ends at the home ranch near the White Mountains.

CHAPTER 23

Wilderness Interlude

Ranger Harlan Johnson, who had spent most of the day fighting a forest fire off beyond Heber, drove up to Mule Springs Corral in his pickup to welcome us to Sitgreaves National Forest, and more particularly to check Rosalio's trail permit. Seeing my cameras, he at once forgot about sheep and began talking photography. Taking pictures was a passion with him; he soon became less interested in the herd's pursuit of the trail than in mine. He had just photographed some elk the day before in connection with the elk census he was taking. I soon became as interested in his elk stories as he was in my camera work.

Most favorable place to observe the animals, Ranger Johnson said, was the Ryan Ranch, just a few miles distant. Rosalio and Pablo would be spending the night near the corral; Johnson suggested that since no one was at Ryan's now, I might want to spend the night there and watch for elk. On a narrow fire trail he drove me through dense forest until we came to an open meadow with ranch buildings at the far end. Frightened by our motor, four elk in one band, and thirteen in another, swept across a pasture, over a fence and into the blackening forest.

"Ratio's just about right in that herd," said Johnson. "Four cows to each bull." I had been so excited that I hadn't even observed their sex, so Johnson went on to explain. "With a four-to-one ratio, and plenty of young coming on, the balance is just about right. If the cows increase too much, we permit more hunting. If they decrease, the hunting's restricted. I take a census each May. Numbers don't matter. What I'm concerned about is the ratio of cows to bulls."

Saying our good-byes, Ranger Johnson headed back to Heber and I went on alone, toward the three log buildings and a shed which made up Ryan's Ranch. I was told that Ryan is a generous soul who is glad to share his place with the wild animals, and would be equally glad to share it with me, although he and I had never met.

I took up my abode in the loft of the barn, which is open almost entirely at one end facing the meadow, and has some boards removed at the other, giving view right into the forest, against which the building is set. Being tired from my climb up the Mogollon Rim, I went to bed almost at once, throwing my blankets near the open end of the loft, so I could see Jupiter rise above the forest, and Caster and Pollux dance above the meadow.

Now it is morning. Twice during the night I looked out onto the meadow, hoping that the elk had returned after their fright from the motor, but was unable to observe any. I wakened at 5:30 A.M., not having Pablo to call me, and was robbed of almost half an hour of daylight. Looking out I saw three elk — all cows — close to the barn, and a calf alone, two-thirds grown, on the far side of the meadow, its tender stubs of horns just starting to develop, like the fuzz on the face of a high school sophomore.

Light was too dim for filming so I settled back in my bedroll — the morning air was just above freezing — and waited. But before the sun's rays penetrated to the meadow, the elk were gone.

This meadow is small, with the intimacy of a New England village common, but with great pines and log structures — instead of church spires and white houses — creating the encircling intimacy. Most of the trees are mature, or at least well along toward the prime of life — one or two hundred years old, some even three hundred, Ranger Johnson told me. Some are mere spindling upstarts — black-barked, wispy, grown close together — of forty or fifty years.

Far down the meadow I saw some elk, large as mules, with nearly an equal bone structure, though not as fat by far. Each animal had a white patch on its rump. Every one of these beautiful creatures must have sat in a flour barrel. Slowly the three which I saw grazed the lower end of the meadow.

134

As I brought them into focus in my camera lens, something in the foreground was blurring my view; a careful check showed it to be two squirrels, each with black stripes running along the back. Like the great elk, they too were grazing, but unlike their huge companions, the squirrels were scampering for their morning meal. One bent his head daintily, that his clippers might be lined up with the stem, and cut a dried dandelion, which he then ate greedily. His cheeks popped in and out like an old man's eating peanuts. I could nearly follow the squirrel's reasoning as he next clipped a tender yellow dandelion, in full bloom. The dried one had been dainty; this would be dessert. But he took one bite and ejected it in evident disgust. In the next few minutes, among many other things, he had clipped three yellow dandelions, taken a bite of each, and spit it away.

But suddenly the squirrel stopped in the middle of a bite, a half-eaten leaf protruding from his mouth as he stood completely motionless.

From down the meadow came a thunderous neighing — a wild, fearful sort of noise. The squirrel dropped everything and ran for the safety of a nearby log. Up from the lower end of the meadow pranced a magnificent black horse, running with high-arched steps. Ranger Johnson had told me to be on the watch for a sight like this; this was a wild horse — a fiery stallion.

He came fearlessly. I had left the shelter of the loft and was out on the ground to watch the squirrel. The stallion saw me — came closer. Then he stopped right out in the middle of the meadow. His great black head was marked with a white star in the forehead. His tail and mane were long and unkempt. The superb beast eyed me with a meanness which was real, then tossed a snort in my direction which cut the morning air and echoed against the forest trees behind me. The squirrel darted from beneath the log and across the intervening dozen feet to his earthen home.

It would be impossible for me to exaggerate the fierceness of that stallion's snort. Three times he did it, in quick succession, and turned half about each time, but always with his eye beading directly on me. I was an intruder here and that snort was for me and no one else. I had heard a softer version of that sound in the night and had not known its origin or meaning.

At intervals for an hour this black creature continued his feud

135

with me — then suddenly hurled himself in his beautiful prance up to the other end of the meadow, vaulted a cattle guard, and disappeared.

So many distractions had held my interest that I'd forgotten completely about the elk. I looked and they were gone, having been for the time almost the least important things in the meadow, even though it was for them I had come.

Going back up to the barn, I found a grandstand seat on a small platform which jutted out from the open end of my loft, and spent an hour or more writing in my journal, always on the alert, however, for wild horses or elk. Instead, right below me — at the edge of the barn — I spotted a porcupine.

He was as large as I have ever seen, the size of a small bear cub, but more cumbersome in his walk than a bear, and a world more ugly; in fact, he reminded me of an ape waddling along on all fours. His small beady eyes surveyed me when I attracted his attention, and there was an excellent chance to survey him in return as he waddled slowly across an arm of the meadow toward the woods. His coloring intrigued me. The long hairs of his body were black, tipped with white at the ends, but his quills were a lustrous golden green. Those showed along the back, but most particularly at the rump. The animal's tail was a beautiful spray of that gold with the greenish sheen.

The porcupine was grazing, nibbling the grass as a sheep does, his head down, plodding along slowly — very slowly. The meadow this day has been a panorama of vegetarians — elk, stallion, squirrels, porcupine. The only carnivorous creatures I have seen today are a robin catching worms, and swallows in pursuit of insects.

I followed along the meadow floor in silence and came to the conclusion that a porcupine, whatever other modes of protection he may have, is not well equipped for hearing and seeing. Although he was grazing in my direction some of the time, and there were no obstructions, I came within a few feet of him before he acknowledged my presence. His back was partially turned and the golden green quills bristled and shimmered in beauty as he tensed every one.

It is fitting that porcupines inhabit Arizona, for they have much in common with the cholla or jumping cactus. Neither porky nor cholla can throw their spines, but in both cases the deadly hooks come off at the least touch and are vicious in their effect.

By this time, the porcupine had worked close to the encircling forest and was just a few feet from a tree. Waddling fast, but not quite at top speed, he made it and started up, climbing swiftly at first, but then hugging the trunk and working his way up past the middle branches. He was almost perfectly concealed and by going slowly did not attract attention. Had I not known his exact location, I could not have spotted him at all.

Only when I observed him in the tree did I begin to comprehend the reason for the coloring I had admired so much. On the meadow it had taken my eye, but it was a camouflage made for the trees — not the meadows — where he could climb for safety. His dark hair was identical in coloring with the black bark of the younger pines, rather than the reddish brown of the giant trees, because these younger ones were easier to scale, and had branches which could be reached more quickly. Large as he was, he could not have hugged one of the 300 year old giants and slid obscurely up. Its open surface would have betrayed him. He matched the trees which were his size.

And those golden green quills of his; the reason for that shading had not occurred to me at all, until now. Up amongst the pine needles, he became just another clump of needles. The sun sifting down through the branches lit up the pine needles and clothed them a golden green; it did the same to the quills of the porcupine. There was no difference. He became part of the tree, part of its bark and its foliage. The "pine" which goes to make up the name of this animal is descriptive and appropriate.

All nature "shades up" as do the sheep, and this is the time for siesta. By 2:00 P.M. — though the air was chilled — I went out into the middle of the empty meadow, leaving my clothes behind, and took a bath in the tiny stream which — though it comes from a spring somewhere in the meadow itself — immediately gains a depth of some three feet, forming a splendid pool into which I could hang my feet from the grassy side.

Few spots along the Mogollon Rim in this section have the solid advantages and charm of Ryan's Ranch. Though I am using the loft of his barn without rent — and have never met him — I feel that in Bill Ryan the meadow has a friend. His house and buildings are simple, blending with the hill and forest against which they are set. Few signs of humanity scar the beauty. If the meadow must be owned, he seems the one to own it. There is another fact which helps bring me to this conclusion — the location of the outside toilet.

Chic Sale immortalized the privy in a book which dwelt almost painfully on every detail of architecture and reminiscence. But he forgot — or never understood — the most important requirement of all. His book never discusses orientation — the way it faces.

The Ryan Ranch outhouse is a fair distance from the other buildings and faces directly away from them — faces a view so charming that it has never had, and will never have, a door. There to the front is a cairn of rocks — a ledge of honeycombed tufa coming out of the forest, with grasses and forest plants pushing up through cracks in its pitted surface, and pines growing above. Immediately over the privy itself extends the branch of a tree which projects a glorious spray of pine needles into the view. Any picture is better when framed, and all this is framed by the doorway — but never by a door. The woodsy cliff provides complete privacy. This is what I have always dreamed of — a "bathroom" with a view.

More elk came during the afternoon and I carefully observed the ratio of cows to bulls — about three-to-one, of those I counted.

Toward dusk, as I walked across the meadow toward the ranger trail, I could dimly see two large shadowy forms — bull elks — working up to the stream for a drink. Off in the woods on the far side came a strange cry, one which I could not identify. As I stepped across the darkness of the meadow, the ground gave under my feet — damp and oozy with the recent rains. A coldness — a strong-scented good cool feeling — lifted from the meadow floor into my face.

Sounds of a motor reached my ear, and I saw lights approaching. Ranger Johnson had come to drive me into Heber.

CHAPTER 24

Cream Puffs and Branding Irons

Heber is a quiet Mormon settlement with a few stores lining the main street. As Ranger Johnson dropped me in front of the town cafe, the last remnants of activities were being clothed in the deepening darkness. A boy and a girl raced their bikes along the road, stirring up and arousing both dust and barking dogs. A young couple were engaged in shy lovemaking, which didn't get quite as far as a kiss. Otherwise, the street was deserted.

Two men from the sawmill were drinking coffee in the cafe and, although it was past supper time, the owner fixed me some toast and eggs. Yes, he said in response to my query, he could make a place for me to sleep in a small room off the kitchen.

He wasn't talkative until, between bites, I mentioned my day at Ryan's Ranch with the elk.

"Elk! We got hundreds of them around here. Did you chance to see any with the Biglar brand?"

"No brands," I laughed. "These were wild fellows."

"I'm serious," he retorted, and by his tone I realized that he was. "You're a stranger around here. Guess you haven't heard about the Biglar Brothers. One of them was just in here couple of hours ago. Too bad you weren't here to meet him."

The Biglar Brothers, I learned in the next half hour, were as legendary in the Sitgreaves Forest as Paul Bunyan was in the early day lumber camps.

But these tales the cafe owner recounted were not legends but facts. (I later confirmed them by visiting the brothers personally and seeing them in action.) One legend concerned elk.

139

Every year, just before the hunting season, the brothers would ride out into the forests and, rodeo fashion, bulldog several elk, brand them with the Biglar iron, then turn them loose. The cafe owner — people called him Jack — told me that hunters, shooting a wild elk and finding it branded, have come back into Heber either unbelieving of what they saw or too scared to tell that they had shot a branded elk.

At breakfast next morning, a small radio, operated by dry cell storage batteries, was puncturing the air with the first formal music I had heard in a month and a half. A sudden impulse struck me. "Want to sell that radio?" I asked Jack.

"Why not?" came his totally unexpected response. Not quite realizing what I had done, I paid him his price along with the rest of my bill. A rancher gave me a ride in his pickup out of town, then by walking less than a mile I came to a place where Ranger Johnson had told me the sheep would be crossing.

Faint, distant bells. A dog bark. The unmistakable bray of burros. Then Pablo on his horse. He and I went along to a forest-fringed cienega and made noon camp. Rosalio soon came with the herd.

I had kept the radio concealed in a gunnysack, courtesy of the Heber Cafe. Several times on the trek Pablo had expressed a yearning for "one of those radio sets," to bring in some music or news at night. The quiet about the campfire has been such a treasure that I could never have brought myself to the point of introducing a radio into the scene, if I had had to make up my mind to buy one at a store. My impulse query of the cafe owner, and his unexpected response, had taken even me by surprise.

"Maybe you'd like this to sing your burros to sleep at night," I said casually to Pablo, and pulled the surprise from its hiding place.

Pablo didn't say a thing — just smiled — and for several minutes examined the set as though he were reading the labels on one of his pails of lard. Then he switched it on. It was tuned to the same Phoenix station which I had heard at the cafe, the only one which came in at all well during the day, in the midst of this forest. The announcer was giving the weather. "Some gusty winds. Clear and sunny everywhere in the state, except for a chance of slight showers north of the Mogollon Rim," came the voice.

140

"Hell, that's here," grinned Pablo broadly. "But how can he know? No rain here." I wondered, guiltily, if Pablo's almanac weather reports, and his coyote forecasts, would be replaced by daily briefings from Phoenix.

Our camp was a cathedral of pines. The noon fire was roaring briskly, for Pablo had built it up to make embers for cooking a pot of beans. The three of us made a triangle about the blaze as we listened intently to the voice coming instantaneously across the wilderness of forests and canyons and deserts over which we and the sheep had been plodding for well over a month. Pablo was using a log for a seat and leaned forward until the cast of the flames brought out the dark reddish-brown in his face until it nearly matched his shirt.

Rosalio lay on his back, feet crossed and a log for a pillow. I was across from him, looking at him through the smoke of the fire. His face was dark, more now like an Indian's than ever, but when he turned and laughed at something the announcer said — and he laughed often — his gleaming teeth were white marble set in dark mosaic.

"It is now 12:30," came the voice from Phoenix, "time for the Betty Crocker Hour." Then a woman's voice took over. Rosalio and I got up and moved closer to the log where Pablo sat. The three of us bent our ears in deep intensity to the set, as the woman's voice explained to us in intricate detail how to make cream puffs.

Pablo was hanging onto each word as she listed the ingredients — eggs, flour, butter and boiling water — and the amounts which would be needed for a dozen puffs. "Place the butter and water in a saucepan and place on the front of the range," she continued. "As soon as the boiling point is reached, add the flour all at once, and stir vigorously." She followed with the directions for making the cream filling, then concluded with a note of caution: "Unless you are having company for dinner, you should divide the recipe so that you get just six puffs."

"Hell," exclaimed Pablo. "We got our burros and my horse, beside Franc-ees. We could eat it all."

That night, around the fire of another camp, we listened to the Green River Boys, the Rio Grande Rangers, and the Montana Girl, from a station in Clint, Texas. We learned the price of potatoes and flour in El Paso. "Too damn high," murmured Pablo.

141

Warnings of frost came from Flagstaff. A barely audible newscast told about the death in an accident of two Mexican men in Springerville, Arizona. Pablo and Rosalio knew them both. "That's bad," said Pablo. Soon he turned the set off.

As I went to sleep, there were the soft sounds of bells of the sheep, and a gentle bray of a burro. Later two coyotes set up a two-way conversation in the distance, and an owl hooted gently from a nearby pine. Wilderness music was taking over once more.

CHAPTER 25

Left-handed Trees

We are in the high forests — at last.

Like the horny-skinned, dust colored, earth-bound caterpillar that emerges from its cocoon as a soaring butterfly, this sheep trail has changed from the tortures of Ramer Canyon and the Rim Country into a palace of pineclad wonders.

It is Sunday, dawn, and the world is at its prayers; the great pines supplicating heaven with raised branches in thanks for the golden sunlight; the grasses, swaying in a bit of religious fervor, with their brows still moist from a morning baptism of dew; the tiny unopened flowers, petals over their eyes, praying silently in their closets as they were taught to do. Over in the opening, a fallen giant ponderosa lies prone on its carpet of needles — perhaps a Moslem among all these Christians — stretching its branches in prostrate praise to Allah.

The forest incense is burning on a thousand piney and cedared altars and the wind — high priest in this reverent world — is bringing the chant and song of the forest morning to all who worship.

The sermons are in stones, but they hold their tongues. The joy of daylight is in the air, which is sermon enough.

As I stop to make notes in my journal, Boots comes flashing through the trees in hot pursuit of a jackrabbit. But the rabbit knows this bit of woodland better than Boots, and dodges and twists to safety. Panting, yet not dejected, the dog returns to the herd. It was a good bit of fun; a bracer for the morning's work.

A lamb comes upon me in the clearing and stops a second —

startled. Finding me harmless, it nips a stalk of gramma grass and holds it in its mouth, nibbling it upward like an old man chewing a toothpick. If I move, it will jump and be off. But I am a frozen statue so the lamb finishes the blade of grass and slowly continues on.

Some great black ants, in their somber capes and jet hoods, are busy about their business — morticians of the forest — slowly returning the dead trees and plants to the soil.

Up in the ceiling of this forest world there are cirrus clouds, drawn out by some high and vagrant wind until they are like taffy at a candy pull. I look again, and they are the flowing tails of wild horses in a race around the world. Some are tails, long and thin; others are flowing manes.

These trees about me are members of a single mighty family. There stands a ponderosa, the oldest tree in sight — hardened elder now, but once just a carefree sapling at the time Washington chopped that other tree back in Fredericksburg. The cherry tree got into history and the ponderosa belongs there too.

There about it stands its family. Every tree for a length and a half is probably its offspring. Some are now nearly full grown and parents themselves; others — grandchildren perhaps — are just shoots in the ground.

I had thought that I could not survive this long without the nourishment of good music. No music of the kind we ordinarily call such, has reached my ears — except in infrequent campfire songs and the few radio sessions — for well over a month. Yet there has been no need for it with me thus far.

This is because I am daily experiencing music in its most basic forms. Natural sounds — the utterings of birds, crickets, animals — are music at its source. A pine tree is a musical instrument in itself; its long tapering needles are the strings of the harp, and the breezes are the musician's fingers which strum the strings. Among these pines there is treetop music, full branch music, then sometimes an orchestration of the entire forest, when not only each branch and tree, but even the dried leaves and needles on the ground, join in a crescendo of sound. There is strong harmony, in a literal sense, in the movements which a wind sets up in the forest.

I think that the music of sight must be an auxiliary to the music of sound. The massed flow of the sheep, the pattern in the

bark of trees and in the wood of the trunks and limbs from which
the bark has fallen away, the swirls in rocks and in the clouds, the
flow of the landscape — these have a musical feel, a rhythm and
harmony and tone. On many parts of this trail, my days and nights
are filled with music.

Thinking that I might have become lost, Rosalio dropped
back to check on me, along with some stray ewes and a lamb.

I explained to him that I wasn't lost — except in wonder at the
beauty of the forest.

"You see left-hand tree?" he asked, and for a moment I wondered if he had somehow caught a look at my journal with its refer-
ence to a tree's branches being raised like hands to heaven.

"Come, I show you," he said.

As I followed his leading, I noticed that he was looking
mainly at the trees — especially the pines — from which the bark
had been stripped, by disease, or lightning, or death. Suddenly he
stopped, at a great stripped giant, and began tracing the "grain" of
the trunk with his herder's staff.

"You see, she left-handed," he observed. "The cracks all go
around in the wrong direction. That one over there" (pointing to
another stripped tree nearby), "she right-handed. Not many trees
left-handed."

For the first time in my life I began to notice that the swirl of
the "grain" was indeed different in different trees. This can best be
observed when the bark is gone. Most swirl upward and around
from left to right. A few have their "grain" or cracks going verti-
cally, straight up and down. And about one in twenty or thirty, I
discovered, is truly "left-handed."

I had just been sensing the beauty of the tree swirls, compar-
ing them to music, but had never observed closely enough to see if
they went to right or to left.

For the rest of the day, I couldn't see the forest for the trees, as
I watched eagerly to find those that were "lefties."

This day has passed all too swiftly. Now it is night, and Pablo
has added to our enjoyment by building for us a giant fire of oak
and pine and cedar. (Although called "cedars" by many herders
and ranchers, these trees are actually junipers.)

It could have been among the Cedars of Lebanon — or some
such setting — that incense was first developed. If there is one

thing which I would capture to be breathed again between four walls some day next winter it would be the smell of these burning cedars.

Now the fire is revealing the quality of Rosalio's eyes. Nearly all herders have gentle eyes. They are farseeing, penetrating, deep — yet warm. They are eyes conditioned by years of constant contact with silent deserts, untouched forests, majestic mountains, arching skies by day, star-spangled skies at night. They are eyes which have spent a lifetime watching out for a thousand children — the sheep; watching for lost lambs, for poison weeds, for animal enemies; eyes which serve.

Eyes bent to the book would not be like those; the herder's eyes read the forest floor in the daytime and the sky at night. Light of the stars close the lids in sleep. Only herders of the sheep can have eyes like these.

CHAPTER 26

Rosalio's "Children"

On Ranger Johnson's map in Heber I had seen how the sheep trail crossed the Sitgreaves National Forest, swung out over private land past Dry Lake, circled the tiny town of Snowflake, then entered Sitgreaves Forest again before crossing the boundary into Fort Apache Indian Reservation. What the ranger pointed out to me in two minutes, by tracing on the map with his pencil, took as many weeks for the herd and us to cover on foot. But here we were at last, camped close to the boundaries of the Reservation.

I thought it was the clash of thunder which wakened me next morning but soon discovered it was just Pablo splitting juniper wood to build up the fire.

A fire was needed. The air at this nearly mile and a half high elevation stung and pinched with its coldness. In the dog's drinking pan, water was frozen a quarter of an inch thick.

Rosalio had to go out immediately to the sheep, for they were astir, so Pablo and I fried our eggs and drank our sugarless coffee. Then Pablo went to hold the sheep while Rosalio ate, after which Rosalio also headed back toward the sheep. I stayed to guard camp and wipe the dishes.

Hearing footsteps behind me, I turned around. Coming daintily into the circle of the camp was a young girl, bonnet on her head and a basket on her arm. The thought flashed through my mind: "What, in heaven's name, is Little Red Ridinghood doing here?"

It proved to be a girl from an isolated ranch nearby. She had come to sell eggs. I bought a dozen, although we already had as many as Pablo could conveniently pack. Prodded by my questions, the girl told me as much about the country as she knew, and as much about herself as I could get out of her. Before Pablo and Rosalio returned, she was gone. Pablo laughed and asked if she had frightened me. "She comes every year," he explained.

A frosty dawn turned into a sunny morning and when we broke camp after shading up at noon, the weather was warm. In the distance we could see a ranch house — probably the home of "Little Red Ridinghood." Soon we came upon an asphalt pavement. This was U.S. Highway 60, the same route we had crossed at the very beginning of the trek, back near Mesa, a month and a half before. This was the only highway, since that other crossing, with traffic that might endanger the sheep. For half a mile Rosalio drove the herd beside it until we found a stretch with an unbroken view in both directions.

"Now we cross," Rosalio told me. "Cars can see us. They have a chance to stop."

Four cars came along. They stopped. The occupants seemed to enjoy this sight of the massed gray-white wool flowing across the black asphalt, although I think they did not realize this was the termination of a flow which had carried that herd in a turbulent river of torment across much of this wild state of Arizona. Those cars heading west would follow Route 60 down to Mesa and Phoenix, and be there in half a day. We had taken a month and a half.

The driver of one of the cars, if he could have had his way, would probably have arrived in Phoenix even sooner. He was plainly irritated at the delay the sheep were causing; he tried to inch his way through, but Rosalio sternly motioned him back.

"Sheepsies don't like it," he explained to me later. "They get scared. They run."

As the last animal crossed the roadway, the car driver blasted his horn violently, churned his wheels, and throttled in high gear down the road toward Phoenix.

"He hurry too much," stated Rosalio simply. "Those other cars, they good. More better that way."

Pablo with his burros and Rosalio with the sheep continued on. I lingered behind at the highway. One of those cars, one of the "good ones," had carried a California license plate. I sat down and wrote a letter to Helen, hoping I might stop some other tourist and ask to have it mailed. Several lumber trucks passed, and a few other

148

cars. None would stop.

Carefully I erected a forked branch by the side of the road-way, and on it hung the sheet of paper containing my letter, with a note on another sheet giving Helen's address and asking if some kind passerby would supply envelope and mail it. To the note, with chewing gum, I attached a quarter for envelope and stamp.

Five days later, Helen received the letter in Eagle Rock, California. We did not know who had provided the Good Samaritan "Special Delivery" service.

But years later I was making a phone call near Chandler to one of the Dobson Brothers, an important sheep owner. "Yes, I remember your name," he said. "I mailed a letter years ago that you left by the roadside."

Late in the day Pablo went ahead but soon came back with a half-sober smile flashing through his whiskers. Usually clean-shaven, he had been growing a beard after topping the Mogollon Rim.

"There's water at Hog Springs," he announced.

I went ahead and came to a glade — all filled with iris and water weeds and spongy underfoot, with tiny white daisies spilled around like confetti tossed at a wedding party.

I followed the iris and the soft footing, which led toward an earthen bank, behind which was a great pool of water stored up from a spring beyond. When that water got into Pablo's kegs and I drank it from a can at night, it was still tinged with red, which is not a good color to taste. But there in the rubied pool, the sight of it was wine to the spirit, for the sheep had been days without water.

The pool was in a saucer of kiln-red earth, bare of grass. The pines which stood on one side were reflected in the polished ruby surface and the trees on the far side laid out their black shadows on the red earth.

The herd came as I had, smelling water first in the iris glade down below. I heard the churning welter of their baas, then the stampede of bells and bodies. Four days without a drink. They broke over the earthen bank, scrambled and slid down and each baa became a gurgle. Those behind swirled about the edges, for there was room for all and they found it.

Now the whole pool was edged in woolly white, like lace on the fringes of a ruby pillow or white carvings on the frame of a picture. They were kneeling worshippers about a sacred pool. When finally the sheep departed — satisfied — there was a baaing of joy. But Rosalio's face was glum. I wondered why.

As we left the pool he told me that it pained him having to bring the sheep to water this near to night. Of course it had to be, for when water comes along once in four days it must be taken on arrival. "But too bad," said Rosalio. "Those lambs especially. They drink so much they be up all night."

Rosalio, although unmarried, is a true parent, with a parent's instincts, so far as his sheep are concerned. As he predicted, the sheep were up, and milling around, nearly all night long.

CHAPTER 27

Symphonies and a Psalm

On June 5, as we were crossing a stream in a cienega, Rosalio said to me: "This is it." At my questioning, he elaborated: "We here. The home ranch. She begins here." We continued on, as though nothing had happened.

To me, I had visioned entering the home ranch as a major climax; in fact, the apex, the successful conclusion of one of the hardest journeys and one of the greatest adventures of my life.

To Rosalio, it was just another day. Throughout the summer, the sheep would have to be moved regularly from one forage area to another of this vast acreage. Throughout the summer there would be cold nights and rainy days. All summer, Rosalio and Pablo would be up at 4:30 or 5:00 A.M. and, in Rosalio's case, often during the night as well, to protect the herd from wild beasts. Then, with approaching winter and snowfall, it would be nearly a two month's trek back through those torturous canyons, forests, and deserts to winter pastures in the Salt River Valley. The bucks — the rams — are introduced into the herd during the summer months and the return trip would be even harder, for the ewes would be heavy with lambs.

The pampered bucks escape the hardships of the trail. They are trucked several hundred miles to reach summer and winter pastures. (See Chapter 4.)

In some pine needles at the edge of the cienega, I found a weatherworn ram's horn, obviously several years old. Bucks are found only at the home ranch; this was a sign which even I could read that we were nearing the end of our journey. I put the ram's horn in my pack as a memento of the trek.

A mile or so farther on, in another enormous cienega, we came upon the owner, Gunnar Thude, awaiting our arrival. "Damn good we're here," remarked Pablo, as he picked a suitable campsite.

Gunnar had brought large sacks of salt, which Pablo and Rosalio spread out in tiny piles throughout the meadow. The sheep were some distance away in the forest, but as Rosalio gave his familiar call of the salting, "Sheepie, Sheepie, Sheepie," they came running almost frantically. He and Pablo both voiced the even more potent call "Ba-a-a-a-," and the sheep crowded and surged about the salt piles. The animals responded with "baas" of their own, which saturated the camp and filled the air. It was a different quality of sound than they made when running to water; the herd seems to have special language for special occasions.

It was impossible to distribute the salt so that every animal could get an equal share. A few of the less fortunate were not able to get any. The animals crowded about the small piles, there was some pushing — but not excessive — and no fighting. Sheep, even in critical situations, are gentle.

With the salting completed, Rosalio, Pablo, and Gunnar made the official count of the herd. Only four were missing — the little lamb which had fallen behind after leaving the sheep bridge, and the lame animals which had purposely been butchered to put them out of their pain.

Rosalio had actually picked up seven strays since the count at the Mule Springs Corral below Heber. He had done what perhaps no other person could have accomplished — brought all his "children" safely through the ordeal of this roughest of sheep trails.

As the final count was announced he smiled, but said nothing. I broke out into gooseflesh, a measure of the depth of my emotions. I was thinking of the simple courage which Rosalio and Pablo had displayed; of the hardships which they had endured in contributing their bit toward the things which made American life comfortable and good. I was breathing a "thank you" that they — and the sheep — and yes, I also — had concluded the journey safely.

My thoughts were carrying back not only along the trail itself, but back through time. Once more, in this cienega, I was

152

pushing back to the very beginning of things.

That counting of the sheep, the results of which had just been announced; it was the necessity for figuring the sheep count, at the beginnings of civilization, that had stirred the first need for mathematics. The ancestors of these sheep in this cienega before me had helped give birth to one of our most lofty sciences.

That ram's horn, which I had found back at the edge of the woods; it was a veritable trumpet of history. It was the blowing of the rams' horns, seven times, by which Joshua signaled the fall of the walls of Jericho. Before that, it was the sounding of the ram's horn — the biblical trumpet — that heralded Moses' ascent of Sinai and the giving of the Ten Commandments. The sounding of the ram's horn summoned warriors to battle and worshippers to the temple. The ram's horn was among the first instruments of music. Literally, the ancestors of these sheep helped give birth to the symphony orchestra. Tubas, horns, trombones — all are amplifications of the horn of the ram.

At night, over this cienega, the constellation of Aries, the Ram, would be riding high in the sky. Aries is the first of the twelve signs of the Zodiac. Sheep were at the very beginnings of things.

Sheep contributed to the making of light with the use of mutton tallow for early candles. Ancient folklore, such as the Legend of the Golden Fleece, has sprung from the herd. The earliest clothes, blankets, simple beds, tents — even the ceilings of the sacred temples — were made of sheepskins.

As these thoughts came to me, as I mentally reviewed the fifty-two days which lay behind me, I seemed to be clothed in antiquity.

I was realizing that this sheep business as it is carried on in Arizona had not changed greatly since Bible times. I was realizing once again that Rosalio, an unassuming Mexican- American herder, had characteristics similar in many ways to the shepherd David. This trail was an Arizona reenactment of David's Twenty-third Psalm.

"The Lord is my shepherd; I shall not want. He maketh me to lie down in green pastures. He leadeth me beside the still waters. " These words evoked a picture of Rosalio's struggles to provide forage and water in the desert.

"He restoreth my soul," summarized the trail's solitude and grandeur and its spiritual effects.

"He leadeth me in the paths of righteousness for his name's sake. Yea, though I walk through the valley of the shadow of death, I will fear no evil, for thou art with me." Only Rosalio's leadership had brought his charges through the shadows.

"Thy rod and thy staff they comfort me." Rosalio with his crooked staff was the very picture of an Old Testament character.

"Thou preparest a table before me in the presence of mine enemies." I visioned the salting of the sheep.

"Thou anointest my head with oil." Often I had seen Rosalio salve wounded sheep with ointment.

"My cup runneth over." The cup was Rosalio's battered hat, holding water for his dogs. Or those overflowing forest-fringed pools of precious water for the sheep.

"Surely goodness and mercy shall follow me all the days of my life, and I will dwell in the house of the Lord forever." Peace and plenty at journey's end was the herd's reward for fifty-two exacting days.

Sheep are unchanging symbols of humanity's arduous trek across the world. If David could return, he would find that he and Rosalio had common bonds of interest and concern.

The entire 52-day trek was like an Arizona reenactment of the 23rd Psalm.
Upper—"He restoreth my soul." Lower—"He leadeth me beside the still waters."

Upper left—"Though I walk through the valley of the shadow of death . . . thou art with me."
Upper right—"Thy rod and thy staff they comfort me." Lower—"My cup runneth over."

"Surely goodness and mercy shall follow me all the days of my life and I will dwell in the house of the Lord forever."

Epilogue

On a rainy night outside the high school of Webster Groves, Missouri, a woman was passing the auditorium entrance where a large bulletin board announced the program being given inside: "SHEEP, STARS, AND SOLITUDE." She stopped to read the words. Then, for some reason which she herself could not explain, she slipped inside and took a vacant seat, just as the lights were dimmed.

I was presenting our color motion picture of Rosalio's trek with the sheep.

After Helen and I had completed editing the film, I had taken it, in personal appearance showings, to lecture halls across America — Constitution Hall in the nation's capital, Field Museum and Orchestra Hall in Chicago, Town Hall and the American Museum of Natural History in New York, Carnegie Hall in Pittsburgh, Kiel Auditorium in St. Louis, The Brooklyn Academy, Pasadena's Civic Auditorium, Shrine Auditorium of Los Angeles, and universities, clubs and schools from San Francisco to Boston.

Webster Groves was one of these stops, and the audience was particularly appreciative. As I left the hall, the woman who had slipped into the showing on impulse touched my arm.

"Mr. Line. I don't know . . . I can't . . ."

She was trying to keep back tears, but forced herself to continue:

"I have to tell you . . . I was going to end it tonight. I was on my way to the river . . . I saw that sign. I don't know why I came in. Those words, I guess — sheep and stars. But the film . . . Rosalio . . . the courage. It gave ME courage again . . . to live."

There was a moment's pause. Then:

"Thank you, Mr. Line. Thank that man."

With those words, before there was a chance for me to make reply or to obtain her name, the woman disappeared into the darkness.

159

It soon became clear to Helen and to me that this record of the sheep trek and Rosalio's quiet heroism was leaving a deep impression on countless men, women, and children. We found that it appealed to all ages.

Officials of the United States Department of State heard about the production and arranged to have a shorter version of it translated into twenty-three foreign languages for worldwide distribution, to show to people everywhere the workaday heroism of a humble American.

A version in Spanish was distributed in Mexico and Spain. Rosalio — via the film — had made his way back to the land of his forebears.

Encyclopaedia Britannica Educational Corporation began distribution of an English language short version which would put it in schools, universities, and libraries of nearly every American state.

We named this shorter version after Rosalio Lucero himself — MORNING STAR — the English translation of his last name. In Rosalio, people were experiencing a strange quality of greatness combined with simplicity — a combination which is rare.

The number of persons to whom I showed the film personally soon reached more than a million, throughout the United States and Canada.

Two people, however, had never seen it — Rosalio and Pablo. We wanted to show it to those two almost more than to anyone else.

Prospects for such a showing seemed remote. They were trail-bound nearly four months of the year. No motion picture projector or electricity were available at the home ranch in the White Mountains, or anywhere along the trail. Chandler, close to the pastures where they tended the herds in winter, had no lecture course.

After much planning, we arranged a charity showing, put on by the Girl Scouts, in Chandler's high school auditorium. Plans were made for Rosalio and Pablo to attend. This was a big moment for us. Helen and I drove out to their tent near the sheep to pick them up.

Rosalio would not come. Pablo was hesitant. There was no question that they wanted to see the film. But — surrounded by a thousand curious eyes?

No.

I should have realized. I had just returned from those hot

crowded lecture halls. Sitting with the herder and campero about their campfire out in the quiet pasture near the herd, I began again to feel the spirit of the trail. I obtained one more insight into the self-effacement of the man who was able to devote his whole life to his sheep.

Next night, with a borrowed projector, in a combined hay barn and toolshed near the herd, before the smallest audience to which it had ever been presented, we showed the film to the two men whose story it portrayed.

As the familiar scenes lit up the small borrowed screen, there would be a chuckle from one or the other, or both. Often they whispered excitedly to each other. Pablo, three or four times, used an endearing Spanish oath of wonder and amazement. On few film occasions in our lives have Helen and I had a more gratifying experience than that showing.

Rosalio expressed his thanks that they had been able to see it here in the barn, rather than before the huge crowd. "More better that way," he said simply.

Pablo grinned and shook his head in wonderment. "That's damn good," he told me. Excitedly they recounted the scenes to each other as we drove them back to their sheep.

Disappearing into the tent, Rosalio emerged with the stout wooden shepherd's staff which he had carried along the Heber-Reno Trail for a dozen years. He had fashioned it himself from a special piece of wood he had found near the summit of the Sierra Ancha. From crook to tip, it was notched with markings of the sheep counts. It was the staff with which he had lovingly hooked untold dozens of his sheep and lambs, that he might treat them for injuries or render first aid. It was the staff on which he had leaned in relaxation when — after getting the herd past some threatening canyon or escarpment — he watched them seek the safety of more pleasant terrain.

"Maybe you like this," he said simply. "Remind you of sheepsies." He handed me the staff.

So that Rosalio and Pablo could see it, we had arranged our presentation of SHEEP, STARS, AND SOLITUDE in Chandler as a charity affair. But no payment which I have ever received for the showing of this film could equal the gift with which we left Chandler that night. This shepherd's staff of the Heber-Reno Stock Trail is among the true treasures of our years producing documentary films about the world.

Several times, in the years that followed, Helen and I met Pablo and Rosalio and the herd at the Salt River sheep bridge. Once each year, somewhere out on the trail, Rosalio would write us a letter.

Years followed swiftly upon one another, as fast as the sheep running single file for water along Borrego Canyon. One spring we arrived several days too late for the sheep crossing at the bridge.

We had not seen Rosalio for a couple of years and longed to make contact once more. So we drove on the Bush Highway up to Sunflower and went in to Sycamore Creek, on a road which had been greatly improved. We hoped to intercept the herd. It had already passed that point.

We tried at Rock Tank and at Tonto Creek. Rosalio and the sheep had either passed, or not yet arrived. Unsuccessful in finding him, we left notes for him with other herders.

Since we were on our way east, we made a side trip to Albuquerque, in New Mexico, and tried to locate his sister. She had moved from the address which we had.

Some young school children passed us on the sidewalk in the Plaza Square. On impulse, Helen asked them: "I wonder if you know of a Mr. Rosalio Lucero. His sister lives here in Old Albuquerque. We want to find her."

Their eyes lighted up. "Rosalio Lucero? Oh, yes. He is the herder. We see a film of him every year in our school. He is a great man."

With the help of the school children we were able to find the sister. The next year, from her, came word of Rosalio's death, on the trail.

"He is a great man."

It is good that school children in Albuquerque know that about Rosalio.

It is good that school children, as well as adults, across the world, may realize that a sheepherder can exhibit the highest qualities of heroism in his everyday work.

Rosalio's is a type of life that broadens the concepts of greatness.

The End.

162

About the
Author

By age twenty-one Francis Line had hiked to every state in the Union and circled the globe, working his way as he went. He is a University of Michigan graduate, magna cum laude, a member of Phi Beta Kappa and The Los Angeles Adventurers' Club.

Line has lectured with his travel-adventure films at Constitution Hall in Washington D.C.; Columbia University, Town Hall, and The American Museum of Natural History in New York; Carnegie Hall in Pittsburgh; Orchestra Hall and Field Museum in Chicago; Shrine Auditorium in Los Angeles; Pasadena Civic Auditorium, and similar halls nationwide. He has also produced educational films which are used in schools, universities and libraries throughout America.

Francis has written for *THE NATIONAL GEOGRAPHIC MAGAZINE, ARIZONA HIGHWAYS, BOYS' LIFE, WIDE WORLD OF ENGLAND* and, along with his wife Helen, is coauthor of the following books: GRAND CANYON LOVE STORY, the narrative of their sixty years of adventure in the Canyon; *MAN WITH A SONG, Major and Minor Notes in the Life of Francis of Assisi;* and *BLUEPRINT FOR LIVING.*

Francis and Helen now celebrate their wedding anniversary every year by hiking the twenty mile round trip to the Grand Canyon's Colorado River.

The Los Angeles Metropolitan YMCA presented the first MARTIN LUTHER KING, JR. Human Dignity Award to the

Lines, in recognition of outstanding service rendered to the youth of Watts, and to the Navajo Indians, on whose reservation the Lines lived for two years as they produced a documentary motion picture there.

Looking Ahead

MORNING STAR, a shortened 35-minute version of Line's film, SHEEP, STARS, AND SOLITUDE, is available for sale or rent as a 16mm color motion picture. This is also for sale as a video cassette. Contact: Encyclopaedia Britannica, 425 N. Michigan Avenue, Chicago, Illinois 60611 or phone their Customer Service: 1-800-621-3900.

There is the prospect that a plaque — to be approved by the Arizona Historical Society — may be erected, hopefully on Tonto National Forest land at the site of the former sheep bridge, to mark the start of the Heber-Reno Stock Trail and to commemorate the herders and camperos who have given so much of their lives to the sheep.

Rosalio, a Mexican-American, exhibited qualities of heroism and steadfastness that can have a valuable impression on Latino students. The author will give substantial quantity discounts on this book — for use in classrooms or libraries — to school systems with considerable numbers of such students in attendance.

The publishers of this book are pledged to produce only works of highest quality and would like to enlist you, the reader, as a partner. If you value the reading of SHEEP, STARS, AND SOLITUDE, your recommendation of it to your friends will aid greatly in its circulation. Copies may be purchased for $8.95, plus one dollar mailing cost. For the following, write to:

Wide Horizons Press, 13 Meadowsweet, Irvine, CA 92715

✓ Information regarding discounts to Latino schools.

✓ Information on other books by Francis Raymond Line.

✓ Names of friends to whom you would like a folder to be sent describing SHEEP, STARS, AND SOLITUDE.